Chemical Hardness

Chemical Hardness

Ralph G. Pearson

University of California,
Santa Barbara

Weinheim · New York · Chichester · Brisbane · Singapore · Toronto

Pappelallee 3,
D-69469 Weinheim, Germany
National 06201 6060
International (+49) 6201 6060
e-mail (for orders and customer service enquiries): sales-books@wiley-vch.de
Visit our Home Page on http://www.wiley-vch.de

Other Editorial Offices
John Wiley & Sons, Inc., 605 Third Avenue,
New York, NY 10158-0012, USA

John Wiley & Sons, Baffins Lane,
Chichester, West Sussex, PO19 1UD, UK

Jacaranda Wiley Ltd, 33 Park Road, Milton,
Queensland 4064, Australia

John Wiley & Sons (Asia) Pte Ltd, 2 Clementi Loop #02-01,
Jin Xing Distripark, Singapore 0512

John Wiley & Sons (Canada) Ltd, 22 Worcester Road,
Rexdale, Ontario M9W 1L1, Canada

Library of Congress Cataloguing-in-Publication Data

Pearson, Ralph G.
Chemical hardness/Ralph G. Pearson.
p. cm.
Includes bibliographical references and index.
ISBN 3-527-29482-1
1. Acid-base chemistry. I. Title.
QD477.P385 1997
546'.24 — dc21 97-11396
CIP

Deutsche Bibliothek Cataloguing-in-Publication Data:

A catalogue record for this book is available from the Deutsche Bibliothek

British Library Cataloguing in Publication Data

A catalogue record for this book is available from the British Library

ISBN 3-527-29482-1

Typeset by Alden Bookset, Oxford
Printed and bound in Great Britain by Bookcraft (Bath) Ltd
This book is printed on acid-free paper responsibly manufactured from sustainable forestation, for which at least two trees are planted for each one used for paper production.

Contents

Preface vii

Abbreviations used in this book ix

1. The HSAB Principle **1**
- Introduction 1
- Order of Hardness 5
- Solvation Energies 10
- Complex Ions in Water 13
- Symbiosis and Anti-symbiosis 14
- Nucleophilic Reactivity 16
- Other Applications 21
- Summary 25
- References 26

2. Density Functional Theory **29**
- Introdution 29
- DFT and the Concepts of Chemistry 31
- Correlation with Molecular Orbital Theory 38
- The Fukui Function and Local Hardness 42
- Chemical Reactivity 43
- Electronegativity Scales 49
- Calculated μ and η Values 55
- References 56

3. Applications of DFT **59**
- Introduction 59
- Rates of Reaction 68
- Reactions of Free Radicals 74
- Aromatic Electrophilic Substitution 78
- Fukui Functions and Atomic Charges 84
- Improving the Energy from an Approximate Wave Function 88
- Solvation Effects 89
- References 95

4. The Principle of Maximum Hardness **99**
- Introduction 99
- The Maximum Hardness Principle 105
- Tests of the PMH 110

The Softness of Excited States 116
Hardness and the Electronic Energy 119
References 123

5. The Solid State **125**
Introduction 125
Bonds in Solids 126
Details of the Model 129
Bands in Solids 137
Insulators and Semiconductors 146
Some Properties of Metals 156
Clusters and Surfaces 162
Recent Applications of Concepts 168
References 170

6. Physical Hardness **175**
Introduction 175
A Definition of Physical Hardness 180
The Principle of Maximum Physical Hardness, PMPH 184
The Hardness of Molecules 191
References 195

Index **197**

Preface

Chemical hardness is an important property of matter, and its usefulness to chemists (and other scientists) is just beginning to be appreciated. I have written this book because I was sure that no one else would write it; a book by another author on the same general subject would undoubtedly be from a different viewpoint. I readily admit to some personal bias in presenting the material and selecting the illustrative examples.

The book is written at a level which requires a background of an undergraduate degree in chemistry, or other science, plus some graduate work or years of experience as a working scientist. Many of the published papers on chemical hardness are highly mathematical. In most cases I have omitted the mathematical details and concentrated on the final results.

I have included several recent developments in science which I believe are very important, and which seem to be related to chemical hardness in a fundamental way. There is much on density functional theory (DFT), including an introduction to the subject. I hope that this will be useful to those curious about, but not conversant with, this remarkable development.

There is also a great deal on the solid state. This is an area of increased interest to chemists. A simplified model of bonding in solids, covering both ionic and covalent bonding, is presented. This model ties together many properties of simple solids, such as cohesive energies, band gaps and work functions. There is even a chapter on physical hardness, what it is (or might be) and its relation to chemical hardness.

There is, of course, much space allotted to certain Hardness Principles, such as the Principle of Maximum Hardness, or the Principle of Hard and Soft Acids and Bases. An attempt is made to show their wide range of useful application, as well as their limitations.

I wish to express my thanks to my many colleagues who have given of their expertise in diverse areas. I must particularly acknowledge the contribution of Robert G. Parr. Without his insight, chemical hardness would still be an example of what is now called "fuzzy logic", useful but incomplete.

Parr's many students, particularly Weitao Yang and Pratim Chattaraj, have also been of great help to me in understanding the ties between chemistry and DFT. I thank Professor John Gilman of UCLA for lessons on physical hardness. At UCSB Professor William Palke has been especially helpful. And finally I wish to thank Phyllis Claudio, who did much more than type the manuscript.

Santa Barbara, California
November 26, 1996

Ralph G. Pearson

Abbreviations Used in This Book

a.u.	atomic unit(s)
bcc	body-centered cubic
BE	bonding energy
bipy	bipyridine
CN	coordination number
DFT	density functional theory
DOS	density-of-states
EEM	electronegativity equalization method
EHT	extended Hückel theory
EN	electronegative, electronegativity
fcc	face-centered cubic
FMO	frontier molecular orbital
FO	frontier orbital
hcp	hexagonal close-packed
HF	Hartree–Fock
HMO	Hückel molecular orbital
HOMO	highest occupied molecular orbital
HSAB	hard and soft acids and bases
IR	infrared
KS	Kohn–Sham
LCAO	linear combination of atomic orbital basis functions
LUMO	lowest unoccupied molecular orbital
M	molar
MCA	methyl cation affinity
MEP	molecular electrostatic potential
MO	molecular orbital
PA	proton affinity
PMH	Principle of Maximum Hardness
PMPH	Principle of Maximum Physical Hardness
PP	pseudopotential
pyr	pyridine
REPE	resonance energy per electron
SCF	self-consistent field
SET	single-electron transfer
SOMO	singly occupied molecular orbital
STM	scanning tunneling microscopy

TS	transition state
UBER	Universal Binding Energy Relation
UV	ultraviolet
vis–UV	visible–ultraviolet

Energy Conversion Factors

	[eV]	[kcal/mol]	[kJ/mol]
1 eV	1	23.0605	96.4853
1 kcal/mol	4.3364×10^{-2}	1	4.184
1 kJ/mol	1.0364×10^{-2}	0.2390	1

1 The HSAB Principle

INTRODUCTION

The concept of chemical, as opposed to physical, hardness appeared in chemistry in 1963.[1] It arose in connection with a study of the generalized acid–base reaction of G. N. Lewis,

$$A + :B = A:B \qquad \Delta H^\circ \tag{1.1}$$

where A is a Lewis acid, or electron acceptor, and B is a base, or electron donor. Since the acid–base complex, A : B, can be an organic molecule, an inorganic molelcule, a complex ion, or anything that is held together by even weak chemical bonds, the scope of Equation (1.1) includes most of chemistry.

Any insight into the properties of A and B that create a strong bond, or a large value of $-\Delta H^\circ$, would be very useful. It was well known that there is no single order of acid strength, or base strength, that would be valid in all cases. "Strength" here is used in the sense of bond strength: that is, a strong acid and a strong base will form a strong coordinate bond. Indeed, it is fortunate that there is no single order, since then most of chemistry would already have been done or would be predictable.

The earliest observations leading to the concept of chemical hardness go back to the time of Berzelius. It was noted that some metals occurred in nature as their sulfide ores, and some as their oxides or carbonates. We can show this more quantitatively by listing the cohesive energies, $-\Delta H^\circ$, for some binary metal oxides and sulfides, MO and MS (Table 1.1).

$$MX(s) = M(g) + X(g) \qquad \Delta H^\circ\,[\mathrm{kcal/mol}] \tag{1.2}$$

The cohesive energy of the oxides is always greater than that of the sulfides. But the amount, Δ, by which it is greater can vary with the metal. Thus Δ is as large as 54 kcal/mol for Mg and as small as 1 kcal/mol for Hg. It is easy to see why Mg and Ca occur as carbonates, and Hg and Cd occur as sulfides. The oxides, of course, would be converted to the carbonates by CO_2 in the air.

More recent foundations for the concept of chemical hardness lie in the works of Chatt[2] and Schwarzenbach.[3] Independently, they showed that metal ions

Table 1.1 Cohesive Energies of Some Binary Metal Oxides and Sulfides

	Mg	Ca	Ni	Zn	Pb	Ca	Cd	Hg
ΔH°, MO [kcal/mol]	239	254	219	174	158	178	148	96
ΔH°, MS [kcal/mol]	185	222	189	147	137	160	132	95
Δ [kcal/mol]	54	32	30	27	21	18	16	1

could be divided into two classes, (a) and (b), depending on the relative affinities for ligands with various donor atoms, in aqueous solution.

Class (a) $N \gg P > As > Sb$	Class (*b*) $N \ll P > As > Sb$
$O \gg S > Se > Te$	$O \ll S > Se \sim Te$
$F > Cl > Br > I$	$F < Cl < Br < I$

Edwards had done something similar even earlier.[4] His classification, however, depended on the proton basicity and ease of oxidation of various ligands. He also made the important step of comparing rates of reaction of various substrates with the same ligands. The existence of two classes of electrophiles was clearly shown.

Metal ions were simply one group of electrophiles. Class (a) metal ions reacted most rapidly, and more strongly, with nucleophiles which were very basic to the proton. Class (b) metal ions reacted most rapidly with nucleophiles that were easly oxidized. Metal ions are also one class of Lewis acids. The ligands with which they form complexes are simply Lewis (and Brönsted) bases.

Thus it would be possible to classify other Lewis acids as class (a) or (b). Because of a shortage of information on formation constants in water, it is necessary to use a variety of other experimental data. Putting the donor atoms of bases in order of increasing electronegativity gives

$$As < P < Se < S \sim I \sim C < Br < Cl < N < O < F$$

The criterion used is that class (a) acids form more stable complexes with the donor atoms to the right, and class (b) acids prefer donor atoms to the left. This is essentially the same criterion used by Schwarzenbach and Chatt. Often the existence or non-existence of related compounds, or complexes, could be used as an indicator. Rates of reaction could also be used by considering the activated complex to be an acid–base complex, A : B.

Table 1.2 shows the result of this classification, as presented initially.[1] It also shows that classes (a) and (b) were renamed as "hard" and "soft", respectively.

There are two reasons for this change in nomenclature. One is that it is often useful to employ comparative terms for two acids, such as "Hg^{2+} is softer than Pb^{2+}". The other comes about as a result of thinking about the fundamental properties of a given acid which made it class (a) or (b). The acceptor

Table 1.2 Classification of Lewis Acids

Class (a)/Hard	Class (b)/Soft
H^+, Li^+, Na^+, K^+	Cu^+, Ag^+, Au^+, Tl^+, Hg^+, Cs^+
Be^{2+}, Mg^{2+}, Ca^{2+}, Sr^{2+}, Sn^{2+}	Pd^{2+}, Cd^{2+}, Pt^{2+}, Hg^{2+}
Al^{3+}, Se^{3+}, Ga^{3+}, In^{3+}, La^{3+}	CH_3Hg^+
Cr^{3+}, Co^{3+}, Fe^{3+}, As^{3+}, Ir^{3+}	Tl^{3+}, $Tl(CH_3)_3$, RH_3
Si^{4+}, Ti^{4+}, Zr^{4+}, Th^{4+}, Pu^{4+}, VO^{2+}	RS^+, RSe^+, RTe^+
UO_2^{2+}, $(CH_3)_2Sn^{2+}$	I^+, Br^+, HO^+, RO^+
$BeMe_2$, BF_3, BCl_3, $B(OR)_3$	I_2, Br_2, INC, etc.
$Al(CH_3)_3$, $Ga(CH_3)_3$, $In(CH_3)_3$	Trinitrobenzene, etc.
RPO_2^+,$ROPO_2^+$	Chloranil, quinones, etc.
RSO_2^+, $ROSO_2^+$, SO_3	Tetracyanoethylene, etc.
I^{7+}, I^{5+}, Cl^{7+}	O, Cl, Br, I, R_3C
R_3C^+, RCO^+, CO_2, NC^+	M^0 (metal atoms)
	Bulk metals
HX (hydrogen-bonding molelcules)	
Borderline	
Fe^{2+}, Co^{2+}, Ni^{2+}, Cu^{2+}, Zn^{2+}, Pb^{2+}	
$B(CH_3)_3$, SO_2, NO^+	

atoms of the first class are usually of high positive charge and small size, with no unpaired electrons in the valence shell. Class (b) acids have acceptor atoms of low positive charge and large size, and often have unshared pairs of electrons in the valence shell. These characteristics meant that class (a) acceptor atoms are not very polarizable, whereas class (b) acceptor atoms are very polarizable.

Since polarizability means deformation of the electron cloud in an electric field, and since things that are easily deformed are soft, this leads to the two classes of acids being called hard and soft, respectively. What one really has in mind is deformation in the presence of other atoms or groups to which bonding was occurring. Thus optical polarizability, although a useful measure of softness, is not quite the correct measure.

Looking at the list of donor atoms for bases given above, it is obvious that polarizability is high on the left side and diminishes as one goes to the right. By the same argument as before, bases which donor atoms such as As, P, Se, S or I are called soft bases. Bases with F, O and N are hard bases. A hard base has an electron cloud that was difficult to deform chemically. Electrons were held tightly, so that loss of an electron is difficult, whereas a soft base is easily deformed and even oxidized. Table 1.3 shows some typical examples.

With this new nomenclature it is possible to make a simple, general statement: "Hard acids prefer to coordinate to hard bases, and soft acids prefer to coordinate soft bases." This is the Principle of Hard and Soft Acids and Bases, or the HSAB Principle.

Table 1.3 Classification of Bases

Hard	Soft
H_2O, OH^-, F^-	R_2S, RSH, RS^-
$CH_3CO_2^-$, PO_4^{3-}, SO_4^{2-}	I^-, SCN^-, $S_2O_3^{2-}$
Cl^-, CO_3^{2-}, ClO_4^-, NO_3^-	R_3P, R_3As, $(RO)_3P$
ROH, RO^-, R_2O	CN^-, RNC, CO
NH_3, RNH_2, N_2H_4	C_2H_4, C_6H_6
	H^-, R^-

Borderline
$C_6H_5NH_2$, C_5H_5N, N_3^-, Br^-, NO_2^-, SO_3^{2-}, N_2

Note that this Principle is simply a restatement of the experimental evidence which led to Table 1.2. It is a condensed statement of a very large amount of chemical information. As such it might be called a law. But this label seems pretentious in view of the lack of a quantitative definition of hardness. HSAB is not a theory, since it does not *explain* variations in the strength of chemical bonds. The word "prefer" in the HSAB Principle implies a rather modest effect. Softness is not the only factor which determines the values of ΔH° in Equation (1.1). There are many examples of very strong bonds between mismatched pairs, such as H_2, formed from hard H^+ and soft H^-. H_2O, OH^- and O^{2-} are all classified as hard bases, but there are great differences in their base strength, by any criterion.

Obviously, all of the factors which determine bond energies must be taken into account, such as the charges and sizes of A and B, the electronegatives of the donor and acceptor atoms, orbital overlaps, and steric repulsions. The HSAB Principle then refers to an additional stabilization of hard–hard or soft–soft pairs, or a destabilization of hard–soft pairs.

For convenience, let us adopt the modern symbols η (Greek eta = h) and σ (Greek sigma = s) for chemical hardness and softness, respectively. The relationship between the two is $\sigma = 1/\eta$. Also, let us lump all other factors determining the strength of bonds formed by A and B into a single factor, S_A or S_B, "intrinsic" strength.[5] Thus S_B is larger for OH^- than for H_2O.

If there were a single order of strengths for acids, and for bases, then S_A and S_B could be defined by equations such as

$$-\Delta H^\circ = S_A S_B \tag{1.3}$$

The HSAB effect then might be added by the modification

$$-\Delta H^\circ = S_A S_B + \sigma_A \sigma_B \tag{1.4}$$

where the σ terms could be positive numbers for soft species and negative numbers for hard species.

One could, of course, attempt to use Equation (1.4) by taking one set of data to assign values of S_A, S_B, σ_A and σ_B to a number of acids and bases, and then using another set of data to check them. Actually something similar to this has already been done by Drago and Wayland, with their well-known four-parameter equation.[6]

$$-\Delta H^\circ = E_A E_B + C_A C_B \tag{1.5}$$

Here E_A and E_B measure ionic bonding and C_A and C_B measure covalent bonding. This is appropriate, since in looking for the reasons underlying the HSAB effect, it is clear that hard–hard interactions are mainly ionic and soft–soft interactions are mainly covalent.[7] But there are many other effects, such as repulsions due to the overlap of filled atomic orbitals on A and B.[8] For this and other reasons, E and C may not be good measures of S and σ. A disadvantage of Equation (1.5) is that different numbers are needed for different environments.

ORDER OF HARDNESS

Actually chemistry is usually not based on Reaction (1.1), but on the exchange reaction

$$\mathrm{A:B + A':B' = A:B' + A':B'} \tag{1.6}$$

The following conclusion can then be drawn, provided A and A′ are acids of the same strength, or B and B′ are bases of the same strength:

$$hs + sh = hh + ss \qquad 0 > \Delta H^\circ \tag{1.7}$$

where h and s are read as the harder and softer of the two acids (bases).

Equation (1.7) is the result of applying Equation (1.5) to the four acid–base complexes contained in reactions (1.6). The strength terms cancel out, and the σ terms give the value of ΔH°. Now it is not practical to demand that $S_A = S_{A'}$ or $S_B = S_{B'}$, but it is possible to ensure that they be comparable in magnitude. This can be done by requiring that A and A′ be acids of the same charge, and of similar size, and the same for B and B′. This minimizes the influence of the $S_A S_B$ terms, and enhances the $\sigma_A \sigma_B$ terms.

Equation (1.7) is based on Equation (1.5) but does not require numbers for its application, only the relative hardnesses of the reactants. But these can be obtained by applying Equation (1.7) to a set of related data. Turning back to cohesive energies, we can assume that the metal ions are all acids of comparable strength. The cohesive energies of all the oxides are larger than those of the sulfides because the oxide ion is a stronger base, due to its smaller size.

Similarly, the value of ΔH° for MgO is larger than that for HgO because Mg^{2+} is smaller than Hg^{2+}. Nevertheless, the strengths are comparable, not grossly different as they would be, say, for Al^{3+} and K^+. Accordingly, Equation (1.7) enables us to write the order of increasing softness as

$$Mg^{2+} < Ca^{2+} < Ni^{2+} < Zn^{2+} < Pb^{2+} < Cu^{2+} < Cd^{2+} < Hg^{2+}$$

using the values of Δ and the reasonable assumption that the sulfidle ion is softer than the oxide ion.

By comparing only acids of the same charge, and bases of the same charge, we have the great advantage that the predictions of Equation (1.7) are not sensitive to the environment.

$$CaS(s) + CuO(s) = CaO(s) + CuS(s) \qquad \Delta H^\circ = -14\,\text{kcal/mol} \tag{1.8}$$

$$CaS(g) + CuO(g) = CaO(g) + CuS(g) \qquad \Delta H^\circ = -30\,\text{kcal/mol} \tag{1.9}$$

Also we see from Table 1.2 that in aqueous solution Cu^{2+} is softer than Ca^{2+}. Indeed, Table 1.2 is in agreement with the ordering above, but with less detail for the various metal ions.

A further check on the hardness ordering can be made using a different set of reference reactions. The bond dissociation energies of fluorides and iodides were used in the earliest attempt to order the metal ions:[9]

$$MF_2(g) = M(g) + 2F(g) \qquad \Delta H^\circ_F \tag{1.10}$$

$$MF_2(g) = M(g) + 2F(g) \qquad \Delta H^\circ_I \tag{1.11}$$

Defining Δ as $\Delta H^\circ_F - \Delta H^\circ_I$, we find the same order of Δ as before, and assuming that I^- is softer than F^-, the same order of increasing softness.

Note that even though we are considering the reactants to be M^{2+} and F^- or I^-, it is possible to use ordinary bond energies because only differences are important. Table 1.4 gives the empirical hardness order for a number of singly charged Lewis acids, based on the bond dissociation energies of fluorides and iodides, D°_F and D°_I.[10] We see that SiH_3^+ is the hardest acid of this group, followed by CF_3^+. In accordance with Table 1.2, Cu^+, Ag^+ and OH^+ are very soft acids.

The same kind of analysis may be made to rank a series of bases in order of increasing softness. Because of the large amount of data, it is convenient to use H^+ and CH_3^+ as the reference acids, with H^+ being the harder. Table 1.5 gives the hardness order for a number of singly charged anions, using $\Delta = D^\circ_H - D^\circ_{CH_3}$ as the criterion. As expected, F^- and OH^- are the hardest bases. Simple carbon ions are much softer, e.g., CH_3^- and $C_6H_5CH_2^-$.

Table 1.4 Empirical Hardness Parameters for Cationic Lewis Acids[a]

Acid	D_F° [kcal]	D_I° [kcal]	Δ [kcal]
SiH_3^+	150	71	79
CF_3^+	130	54	76
CH_3CO^+	120	50	70
HCO^+	122	52	70
H^+	136	71	65
$C_6H_5^+$	124	64	60
$C_2H_5^+$	119	63	56
t-$C_4H_9^+$	108	50	58
i-$C_3H_7^+$	107	53	54
$C_2H_5^+$	107	53	54
CH_3^+	109	56	53
$C_3H_5^+$	98	44	54
c-$C_3H_5^+$	111	59	52
Li^+	137	82	55
Na^+	123	69	54
Tl^+	105	64	41
CN^+	112	73	39
NO^+	56	20	36
Cs^+	118	82	36
I^+	67	36	31
Cu^+	102	75	27
Ag^+	87	61	26
HO^+	56	52	4

[a] After Reference 10.

The cyanide ion is an ambident base. The carbon end is seen to be much softer than the nitrogen end, as would be predicted. The greater strength of the H–CN bond, compared with H–NC, means that the carbon end is much more basic than the nitrogen end. This would be true even in solution, since the ions are the same for both acids. As a result, binding to carbon will be more common than binding to nitrogen, even for hard acids.

Tables 1.4 and 1.5 contain acids or bases with a wide range of acceptor or donor atoms. In such cases, use of different references will not give identical orderings. Usually the variations are not very great, being shifts of two soft, or two hard, acids or bases with respect to each other.[10,11]

The most reliable (i.e., transferable) results are obtained if two similar acids or bases are used as references. Thus OH^- and SH^- will give the same ordering as F^- and I^-, but H^- and F^- will give orders that are quite different: the alkali metal cations will be much harder than in Table 1.4. Such changes in order are evidence for special bonding effects in certain A : B combinations.

Table 1.5 Empirical Hardness Parameters for Anionic Bases[(a)]

Base	D°_{H} [kcal]	$D^\circ_{CH_3}$ [kcal]	Δ [kcal]
F^-	136	109	27
Cl^-	103	85	19
Br^-	88	70	18
I^-	71	56	15
OH^-	119	92	27
SH^-	91	74	17
SeH^-	79	67	12
NH_2^-	107	85	22
PH_2^-	87	76	11
AsH_2^-	75	63	12
$CH_3CO_2^-$	106	83	23
$C_6H_5O^-$	87	64	23
NO_3^-	102	80	22
CH_3O^-	104	83	21
HO_2^-	88	69	19
ONO^-	78	60	18
NO_2^-	<78	61	<17
NCS^-	96	77	19
$C_6H_5NH^-$	88	71	17
$n\text{-}C_3H_7S^-$	87	72	15
$C_6H_5S^-$	83	69	14
CH_3^-	105	90	15
SiH_3^-	91	89	2
GeH_3^-	87	83	4
$C_6H_5CH_2^-$	88	72	16
$NCCH_2^-$	93	81	12
$CH_3COCH_2^-$	98	86	12
$C_3H_5^-$	86	74	12
$C_6H_5^-$	111	100	11
$C_2H_3^-$	110	105	5
HC_2^-	130	122	8
CH_3CO^-	96	91	5
CF_3^-	106	101	5
CN^-	124	122	2
NC^-	110	98	12
H^-	104	105	−1

[(a)] After Reference 1

In the case of H^-, the special effect is the total absence of π-bonding or π-antibonding. Hydride ion is a pure σ-donor, compared with iodide ion. Figure 1.1 shows the π-interaction of a d (or p) orbital on an acceptor atom, and a p (or d) orbital on a donor atom. If the d orbital is filled, and the p orbital is empty, there is a stabilizing effect. This would be the case in a soft–soft A : B complex. If the

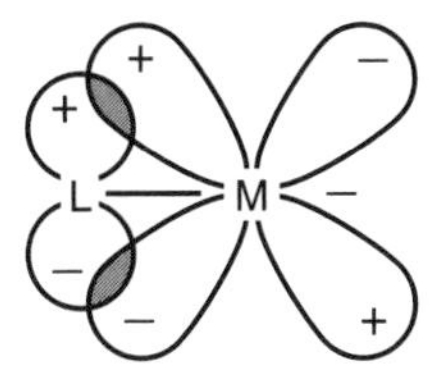

Figure 1.1 A p orbital on a ligand atom and a d orbital on a metal atom. The plus and minus signs refer to the mathematical signs, or phases, of the orbitals in different regions of space.

d orbital is empty, and the p orbital is filled, there would also be a stabilization, as in a hard–hard combination. If both the d and p orbitals are filled, there would be a destabilization, as in a hard–soft combination.

These π-bonding effects are part of the theory of the HSAB Principle. We can also imagine that London dispersion energies between atoms or groups in an A : B complex could stabilize it. Since these dispersion energies, or van der Waals energies, depend on the product of the polarizabilities of the two groups, soft–soft combinations would be stabilized in this way. The hydride ion is very polarizable, and its softness depends on this factor, presumably.

Because of the absence of π-orbital effects, the bond strength of HX, where X is any element, depends almost entirely on the electronegativity of X. D° ranges from 42 kcal for CsH to 136 kcal for HF, for the Main-Group elements. For the

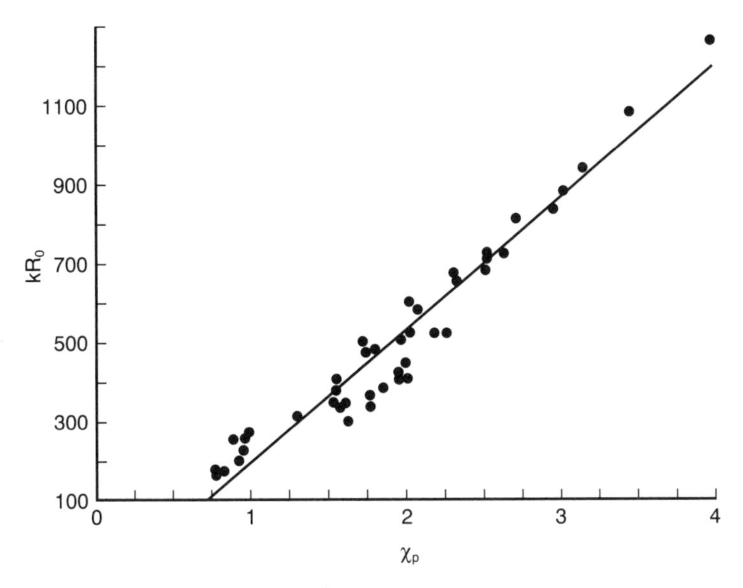

Figure 1.2 Plot of force kR_0 [kcal/mol Å] against the Pauling electronegativity of X, for HX molecules. Reprinted from R.G. Pearson, J. Mol. Struct. (Theochem.), *300*, 519 (1993), with kind permission of Elsevier Science – NL, Sara Burgerhartstraat 25, 1055 KV Amsterdam, The Netherlands.

Table 1.6 Comparison of Na^+ and Cu^+ as Lewis Acids

Reaction	ΔH° [kcal/mol]
$NaH(g)=Na^+(g)+H^-(g)$	151
$NaF(g)=Na^+(g)+F^-(g)$	163
$CuH(g)=Cu^+(g)+H^-(g)$	222
$CuF(g)=Cu^+(g)+F^-(g)$	202
$H_2(g)=H^+(g)+H^-(g)$	403
$HF(g)=H^+(g)+F^-(g)$	371

transition metals, the range is more variable because of the necessity to promote to the valence state (half-empty s orbital).[13] Promotion energies are also important for the hydrides of Groups 2 and 12.

Figure 1.2 shows kR_0 plotted against the Pauling electronegativity, χ_p, for all the diatomic hydrides where the data is available. The quantity kR_0, where k is the force constant and R_0 the internuclear distance, is simply the force that the atom X exerts on the H atom.[13] This force is a good measure of what the bond energy would be without the interference of promotion energies.

As expected, H^+ is also a rather special Lewis acid because of the absence of filled inner shells. Consider the reactions in Table 1.6. The results show that Cu^+ is a "stronger" acid than Na^+, though the "strengths" are comparable. This accords with copper being more electronegative than sodium. The numbers also show clearly that Cu^+ is softer than Na^+. The proton is seen to be a very strong acid because of its small size and ability to burrow into the electron cloud of bases. There is no destabilization with the soft hydride ion in H_2. Note that hydride ion is a strong base, comparable with fluoride ion.

SOLVATION ENERGIES

As mentioned earlier, a method for classifying Lewis acids is based on relative equilibrium constants *in water* for the formation of complexes. Hydration energies must play an important role. As Table 1.4 shows, F^- forms a stronger bond to all cationic acids in the gas phase than I^- does. But because the hydration energy of F^- is 50 kcal greater than for I^-, inversions in ΔH°, and in formation constants, can occur for acids like Ag^+, but not for CH_3CO^+.

For solvents less polar than water, the difference in the heats of solvation will usually be less than 50 kcal, and inversions in order will not occur. Thus, even for Rh(I) and Pt(II) the order Cl > Br > I is found in most non-aqueous solvents. But this is not a violation of the HSAB Principle.

Table 1.7 Bond Dissociation Energies for $M:B^+ = M^+ + :B$ in the Gas Phase[(a)]

$M:B^+$	ΔH° [kcal/mol]	$M:B^+$	ΔH°
$Ag(H_2O)^+$	29.2	$Li(H_2O)^+$	36
$Ag(H_2S)^+$	32.4	$Li(H_2S)^+$	25
$Ag(H_2Se)^+$	34.8	$Li(H_2Se)^+$	24.5
$Ag(NH_3)^+$	43.3	$Li(NH_3)^+$	41
$Ag(PH_3)^+$	40.5	$Li(PH_3)^+$	27
$Ag(AsH_3)^+$	25.4	$Li(AsH_3)^+$	23
$Ag(HF)^+$	16.3	$Li(HF)^+$	23.5
$Ag(HCl)^+$	18.2	$Li(HCl)^+$	18.0
$Ag(HBr)^+$	21.0	$Li(HBr)^+$	17.0

[(a)] Data from Reference 14.

For neutral molecules acting as bases, the effect of solvents would be similar to that of ions, but smaller in magnitude.[1] That is, a hard acid solvent like water would deactivate other hard molecules more than related soft molecules. Table 1.7 lists the calculated bond dissociation energies for a number of complexes in the gas phase

$$M:B^+(g) = M^+(g) + :B(g) \qquad \Delta H^\circ \tag{1.12}$$

The numbers are theoretical ones, but the method of calculation used gives very good agreement with experiment, where numerical results have been obtained.[15]

It might be thought that the slightly stronger bonding in $Ag(NH_3)^+$ than in $Ag(PH_3)^+$ is a failure of the HSAB Principle, since Ag^+ is soft. But this overlooks the requirement that data in aqueous solution are to be used. Since the heats of hydration of NH_3 and PH_3 are -8.1 kcal/mol and -3.6 kcal/mol, the difference is enough to overcome the small difference in bond energies for Ag^+ (2.8 kcal/mol), but not enough for Li^+ (19 kcal-mol).

The HSAB Principle can also be tested by applying equation (1.7):

$$Ag(NH_3)^+(g) + Li(PH_3)^+(g) = Ag(PH_3)^+(g) + Li(NH_3)^+(g) \qquad \Delta H^\circ = -16\,\text{kcal/mol} \tag{1.13}$$

This test works for all the examples in Table 1.2, and also for similar data for Na^+ compared with Ag^+.[16]

Table 1.8 has some data on the proton affinities (PA) of a number of neutral molecules:

$$B(g) + H^+(g) = BH^+(g) \qquad -\Delta H^\circ = PA \tag{1.14}$$

Table 1.8 Gas-phase Proton Affinities of Some Neutral Bases[a]

Base	PA [kcal/mol]
HF	117
H_2O	167
NH_3	205
CH_4	121
HCl	135
H_2S	170
PH_3	189
HBr	139
H_2Se	171
AsH_3	179

[a] Data from Reference 16.

The results for H_2O, H_2S and H_2Se are surprising and would not have been anticipated in 1963, when few data on PAs were known. The strength of bonding increases with the size of the donor atom, even though the base strength is not expected to increase, and the softness does increase. The opposite trend should have been observed, as it is for NH_3, PH_3 and AsH_3. The hydrogen halides show an even greater reversal.

Neither Equation (1.4) nor Equation (1.7) can explain these results. It is a gas-phase phenomenon, since it vanishes in solution, where the expected orders are found for H_2O, H_2S and H_2Se.[16] The simplest explanation is that the inverted order is the result of the classical charging energy for a sphere. This energy (the Born charging energy) is given by

$$\Delta E = q^2/2\varepsilon R \tag{1.15}$$

where q is the charge, R the radius of the sphere, and ε the dielectric constant of the medium.

In more detail, since the proton in Equation (1.14) bears a positive charge, the energy increase results from the repulsion of the proton by the shielded nuclear charges of B. These are largest for a short bond to the proton, as we have in H_3O^+, and are lower for H_3S^+ and H_3Se^+. Also the nuclear repulsions decrease as charge is removed from the nucleus to the outer parts of the molecule, as in going from H_2F^+ to NH_4^+. Since Equation (1.15) depends inversely on R, the overall effect is to make small molecules reluctant to add protons in the gas phase.

This is unfortunate, since there is a great deal of data available on the methyl cation affinities (MCA) of neutral molecules[17]

$$B(g) + CH_3^+(g) = BCH_3^+(g) \qquad -\Delta H^\circ = MCA \tag{1.16}$$

Comparison of MCAs with PAs, as in Equation (1.14), could lead to a hardness ordering of neutral molecules, just as was done for anions in Table 1.5. But small molecules, which are usually hard, will appear too soft because of the Born charging energies. This problem does not arise for anionic bases.

COMPLEX IONS IN WATER

Complex formation between metal ions and ligands in aqueous solution has always been of great interest. This interest is enhanced by the role of metal ions in biology. The first use of the concept of chemical hardness was to explain complex ion formation in water. However, it turns out that this is not so easy as one might expect.

There is a great deal of data available, not only on equilibrium constants, but also on heats of reaction between metal ions and various ligands.[18] For various practical reasons, it is difficult to find data that can be used even in a comparative sense. Solvation energies are one obvious problem. Only for restricted cases can we use Equation (1.7).

Nevertheless, some useful generalizations may be made. Consider the typical reaction for the formation of a complex in water (charges are omitted for simplicity):

$$ML_n(H_2O)(aq) + Y(aq) = ML_nY(aq) + H_2O(l) \qquad (1.17)$$

There are four interactions in Equation (1.17) which are acid–base in character. One is the interaction of ML_n with H_2O, the second is the ML_n interaction with Y, the third is the interaction of Y with the solvent, and the fourth is the interaction of water with itself, a constant factor. The solvation energies of the large complex ions, $ML_n(H_2O)$ and ML_nY, are governed primarily by the Born equation.

If ML_n is hard and Y is soft, or vice versa, we can expect ΔH° to be positive, or only a small negative number. Therefore, complexation will not be favorable. Suppose ML_n is hard and Y is hard. Then their interaction will be favorable. But H_2O is a hard molecule, both as an acid and a base. Therefore Y(aq) and $ML_n(H_2O)$ are also stabilized. Overall, only a small negative value for ΔH° can be hoped for, and moderately stable complexes.

But if ML_n is soft and Y is soft, then everything works in favor of a large negative value for ΔH°, and very stable complexes. Ahrland has made a detailed study of the available data and has found a remarkable agreement with the above predictions.[19] Hard acids rarely form complexes with soft bases, and hard bases do not form very stable complexes with soft acids, except for strong bases such as OH^-.

Hard acids form only moderately stable complexes, even with hard bases, and ΔH° is close to zero. Soft acids and soft bases usually form very stable complexes in aqueous solution, and ΔH° is then a large negative number. For a soft reference acid, such as Pd^{2+} or CH_3Hg^+, the order of increasing values of $-\Delta H^\circ$ is $F^- \ll Cl^- < Br^- \sim N_3^- < I^- < SCN^- < RS^- < CN^-$.[18–20] For neutral ligands the order is $H_2O < NH_3 <$ thiourea $< PR_3$. Note that the order for anions is not exactly the same as in Table 1.5 because of different hydration energies.

There are not enough comparable data to put many metal ions in rank order, but the following orders of decreasing values of $-\Delta H^\circ$ can be established: $Hg^{2+} \gg Cd^{2+} > Zn^{2+}$; $Pt^{2+} \sim Pd^{2+} > Ni^{2+}$. Also the Irving–Williams order[21] for formation constants of complexes, $V^{2+} < Cr^{2+} > Mn^{2+} < Fe^{2+} < Co^{2+} < Ni^{2+} < Cu^{2+} > Zn^{2+}$, follows the experimental values of η for these metal ions. As expected, the magnitude of the changes in $\log K_{eq}$ increase with the polarizability of the ligand donor atoms.[22]

There is an excellent recent review of the role of metal ions in biology by Hancock and Martell.[23] The discussion is in terms of HSAB, but for quantitative work the most accurate treatment is by an equation related to Equation (1.5):

$$\log K_1 = E'_A E'_B + C'_A C'_B - D_A D_B \tag{1.18}$$

K_1 is the equilibrium constant (in H_2O, at 25 °C) for Reaction (1.1), and the E' and C' parameters are analogous to those in Equation (1.5), but are empirical numbers from data in water.[24]

The D parameters are necessary to account for steric effects. The last term in Equation (1.19) can be neglected for large cations, but is important for small cations. It emphasizes another important difference between reactions in solution and in the gas phase. In solution the Lewis acid is multicoordinate, and has a number of water molecules attached to it. The steric effects arise from clashes of the ligand with these water molecules. A large number of E', C' and D parameters are now available, and can be used to estimate formation constants for complexes between metal ions and the common ligands.

SYMBIOSIS AND ANTI-SYMBIOSIS

In Table 1.3, BF_3 is listed as a hard acid but BH_3, which also contains B(III), is considered a soft acid. For example,[25]

$$BH_3:NH_3 + BF_3:CO = BH_3:CO + BF_3:NH_3 \qquad \Delta H^\circ = -12\,\text{kcal/mol} \tag{1.19}$$

Also, in Table 1.4, CF_3^+ is shown as a harder acid than CH_3^+. These are examples of a very general phenomenon, first noted by Jorgensen and called by him the "symbiotic" effect.[26] Soft bases attached to the same central acceptor atom make it a soft acid, and hard bases make it a hard acid. In coordination chemistry, symbiosis explains why some ligands, such as CN^- or phenanthroline, make a metal ion form strong complexes with other soft ligands, whereas F^- and H_2O favor the bonding of other hard ligands.[27]

The symbiotic effect is also common in organic chemistry, but here it has been called the clustering, anomeric, or geminal effect.[28,29] Clustering refers to the stabilization caused by adding several substitutents to the same carbon atom. Some extreme examples are shown by Reactions (1.20) and (1.21), in which the number of bonds of each kind is preserved.

$$4CH_3F(g) = 3CH_4(g) + CF_4(g) \qquad \Delta H^\circ = -63\,\text{kcal/mol} \tag{1.20}$$

$$4CH_3OCH_3(g) = 3CH_4(g) + C(OCH_3)_4(g) \qquad \Delta H^\circ = -52\,\text{kcal/mol} \tag{1.21}$$

In Reaction (1.20) the comparison is between F^- and H^-, the hardest and softest bases in Table 1.5, and the effect is at a maximum.

Elements other than carbon may be influenced, for example[29,30]

$$4SiH_3F(g) = 3SiH_4(g) + SiF_4(g) \qquad \Delta H^\circ = -23\,\text{kcal/mol} \tag{1.22}$$

Comparing Reactions (1.22) with (1.21), it appears that silicon is less affected than carbon by clustering. But consider the reaction

$$SiF_3H(g) + CF_4(g) = SiF_4(g) + CF_3H(g) \qquad \Delta H^\circ = -37\,\text{kcal/mol} \tag{1.23}$$

Clearly SiF_3^+ is much harder than CF_3^+, just as SiH_3^+ is harder than CH_3^+. An extreme example of the difference between the two elements is given by

$$\begin{aligned} &C(OCH_3)_4(g) + SiH_4(g) \\ &\quad = CH_4(g) + Si(OCH_3)_4(g) \qquad \Delta H^\circ = 144\,\text{kcal/mol} \end{aligned} \tag{1.24}$$

This illustrates the great affinity of silicon for binding to oxygen donors.

Organic chemists usually explain the anomeric effect by double-bond–no-bond resonance.[29] This is consistent with the observation that F^-, OR^- and NR_2^- bases give the largest effects. However, the HSAB principle provides a much simpler explanation. The acceptor atom in BF_3, CF_3^+ and SiF_3^+ is much more positive than in BH_3, CH_3 and SiH_3^+. Since a high positive charge enhances hardness (Table 1.3), we now have acids which bind better to hard bases, such as F^-, than to soft bases such as H^-.

In spite of this simple and logical explanation, there are many cases of anti-symbiotic behavior for the soft metal ions of the second and third transition series.[31] That is, soft bases attached to a soft acid center, such as Ir(I), Hg(II), Pt(II) and Au(III), can lower the affinity for another soft base. But this only occurs for a coordination site *trans* to the original ligand. The rule is that two soft ligands in mutual *trans* positions will have a destabilizing effect on each other, when attached to a soft metal ion.[31]

This rule explains a host of experimental observations. For example, $(CH_3)_2Hg$ is readily cleaved by dilute acid to form linear $CH_3Hg(H_2O)^+$. But this cation resists cleavage, even by strong mineral acids. There is a simple explanation based on the theory of the *trans* effect.[32] This effect occurs only for soft metal ions, and ligands of high *trans* effect are always soft ligands.

The theory of the effect postulates that the bonding in such cases is largely covalent. Ligands in *trans* positions compete for the same orbitals to form covalent bonds (σ or π). Therefore it is advantageous to have a hard ligand, such as H_2O or OH^-, *trans* to each soft ligand already attached.[33] Anti-symbiosis should be minor in tetrahedral complexes, or when the acceptor atom is hard.

NUCLEOPHILIC REACTIVITY

Early evidence for the HSAB Principle came from studies of nucleophilic reactivity series towards different substrates, or electrophiles.[34] Some electrophiles, such as H^+, in proton transfer reactions or CH_3CO^+ substitution reactions of esters, reacted rapidly with bases that were strong bases towards the proton. Other electrophiles, such as Pt(II) or RO^+, reacted rapidly with polarizable bases, and were indifferent to proton basicity.

It soon became clear that these two classes should be called hard and soft electrophiles, respectively. Since the terms nucleophile and electrophile refer to rates of reaction, by definition, the acid–base reaction involved is

$$B' + A:B = B':A:B \rightarrow B':A + B \tag{1.25}$$

The rate then depends on the stability of the activated complex B′ : A : B. This, in turn, depends mainly on the compatiblity of B′ and A, if relative rates for a series of nucleophiles, B′, are compared.

Figure 1.3 shows the rate constants for the hydrolysis of *p*-nitrophenyl acetate, catalyzed by the attacks of bases on the carbonyl carbon.[35] There is a rough proportionality between $\log k$ and pK_a. The scatter is expected when a number of bases of different natures are compared. While F^- is fairly reactive, Cl^-, Br^- and I^- show no measurable reactivity.

In contrast, Figure 1.4 shows rate data for many nucleophiles reacting with a pyridine (Pyr) complex of Pt(II) in methanol at 25 °C.[36]

$$B + trans\text{-}Pt(pyr)_2Cl_2 \xrightarrow{k} trans\text{-}Pt(pyr)_2ClB^+ + Cl^- \tag{1.26}$$

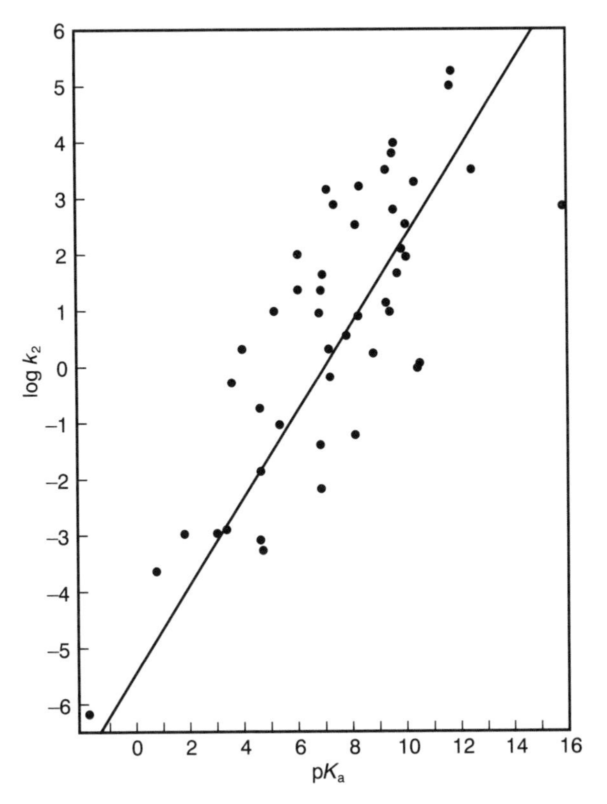

Figure 1.3 Rate constants for the reactions of nucleophiles in aqueous solution at 25 °C, plotted against the basicities of the attacking reagents. After Reference 35.

The rate constants are shown as a parameter, n, defined as

$$n = \log(k/k_s) \tag{1.27}$$

where k_s is the rate constant for solvolysis. The plot is that of n against the pK_a in water, but using pK_a values in methanol, when known, makes no difference. Clearly there is no dependence of n on the normal proton basicity. In fact CH_3O^-, the strongest base possible in methanol, has no detectable reaction with *trans*-$Pt(pyr)_2Cl_2$. The same is true for F^-, but Br^- and I^- are good reagents.

An important substrate for nucleophilic reactivity is methyl iodide. This serves as a model for substitution reactions at tetrahedral carbon in general, the S_N2 reaction

$$B + CH_3I \xrightarrow{k} CH_3B^+ + I^- \tag{1.28}$$

The greatest amount of rate data is in methanol at 25 °C.[36] The values of n range from $n = 0$ for the solvent, to $n = 10.7$ for $C_6H_5Se^-$. Some organometallic

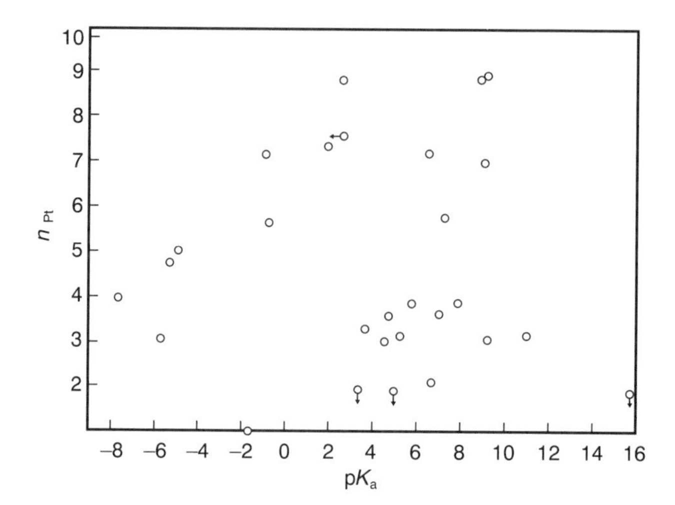

Figure 1.4 Plot of n_{pt}, in CH_3OH, against pK_a in H_2O. After Reference 35.

anions, such as $Fe(CO)_4^{2-}$ and $(C_5H_5)Fe(CO)_2^-$, react even faster.[37] A plot of n against pK_a shows behavior intermediate between that in Figures 1.3 and 1.4. Both CH_3O^- and I^- are good reagents. The best reagents are all soft bases, and the overall picture is that CH_3^+ is a moderately soft electrophile. This is consistent with Tables 1.3 and 1.4.

The values of n cannot be used as an order of increasing softness, since that would ignore the "intrinsic strength" factor. To emphasize the strength factor, it is useful to examine the equilibrium constant for Reaction 1.28. For anionic bases, X^-,

$$X^-(aq) + CH_3I(aq) = CH_3X(aq) + I^-(aq) \qquad -\Delta G^\circ \tag{1.29}$$

There are enough data available to calculate ΔG° for a large number of anion bases, and a smaller number of neutral bases, B.[37]

The order of decreasing values of ΔG° is $H_2O > CH_3OH > Br^- \sim NO_3^- > I^- > F^- > Cl^- > SCN^- > CH_3COO^- > (CH_3)_2S > NO_2^- > C_6H_5O^- > (CH_3)_3As > NH_3 > C_6H_5S^- > n\text{-}C_3H_7S^- > CH_3O^- > (CH_3)_3P > CN^- \gg NH_2^- \sim CH_2CN^- \gg H^- > CH_3^- > C_6H_5^-$. The range of equilibrium constants is 60 powers of 10. The order is that of decreasing thermodynamic leaving-group ability. It is in good agreement with the kinetic leaving-group ability, the nucleofugality, when known.

The next step is to compare n with ΔG°. That is, to look for a linear free energy relationship. This is usually expressed as[38]

$$\Delta G^\ddagger = \alpha \Delta G^\circ + \text{constant} \tag{1.30}$$

where $\Delta G^\ddagger$ is the free energy of activation. Over a wide range of ΔG°, α need not remain constant, but it must change slowly and continuously. Figure 1.5 shows a

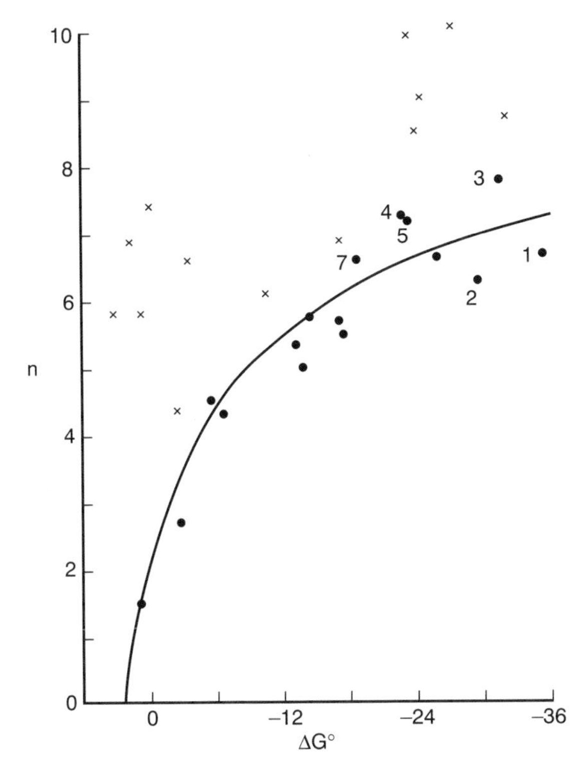

Figure 1.5 Plot of n against ΔG° [kcal/mol] in water at 25 °C, for the reaction of CH_3I with nucleophiles. Circles are for first-row donors (F, O, N, C) and crosses for heavier donors. Key: 1, CN^-; 2, CH_3O^-; 3, HO_2^-; 4, $(C_2H_5)_3N$; 5, piperidine; 6, pyrollidine; 7, N_2H_4. Reproduced with permission from R.G. Pearson, *J. Org. Chem.*, **52**, 2131 (1987). © 1987 American Chemical Society.

plot of n, which is proportional to $\Delta G^\ddagger$, against ΔG° for the cases where n is known. First-row donors (F, O, N, C) are shown as circles and higher-row donors as crosses. The former all lie close to the smooth curve that has been drawn. The latter all lie above it.

The curve acts as a reference line, but it also shows that the slope, α, in Equation (1.30) has a value near one for ΔG° near zero, and approaches zero as ΔG° becomes a large negative number. Such behavior is an expected consequence of the Bell–Evans–Polanyi–Leffler–Hammond principle.[39] It corresponds to a late transition state for the more difficult reactions and a progressively earlier transition state for more exergonic reactions. However, the curve appears to level off at an n value near 8, which corresponds to a second-order rate constant of about 1 M^{-1} s^{-1} at 25 °C, far from the diffusion-controlled limit.

Recent developments, both experimental[40] and theoretical in nature,[41] help explain this behavior. The energy barrier to reaction consists of a part that exists in the gas phase and in solution, and a solvation part found only in solution. The gas-phase barrier exists because five groups attached to carbon in the transition state are at a higher energy than four. The solvation barrier results because the nucleophile must be partly desolvated before it can react, and the transition state (TS) must be solvated. For the reaction of Cl^- with CH_3Cl, the two parts are of comparable magnitude in water, each being about 14 kcal/mol. Desolvation of Cl^- requires more energy than is returned in solvating the TS.

For neutral nucleophiles the situation is somewhat different. The gas-phase energy barrier is very large, so that reactions such as (1.28) simply do not occur. For example, when B is NH_3 we have

$$NH_3(g) + CH_3I(g) = CH_3NH_3^+(g) + I^-(g) \qquad \Delta G^\circ(g) = 108\ \text{kcal/mol} \quad (1.31)$$

Obviously only strong solvation of the polar TS allows Reaction (1.21) to go with a measureable rate in solution.

The estimation of the gas-phase energy barrier is difficult for anions, even when the rate constants are available. However, we can be reasonably sure that this barrier gets smaller as ΔG°(g) becomes a more negative number. For OH^-, CH_3O^-, HO_2^- and CN^-, ΔG°(g) is −64, −57, −62, and−48 kcal/mol, respectively. It is likely in these cases that the gas-phase barrier has become negligible, and only the solvation barrier remains. This would account for the leveling off of the rate constants shown in Figure 1.5. The limiting rate constant corresponds to a value of $\Delta G^\ddagger$ of about 20 kcal/mol, which is reasonable. The scatter of the n values for these four anions would then be due to variations in the solvation energy barrier. For neutral nucleophiles, the gas-phase barrier can become smaller, but it can never become negligible. Therefore no leveling off will occur. The n values for such nuclelophiles, in fact, continue to rise as ΔG° becomes more negative in Figure 1.5.

We turn next to the crosses in Figure 1.5 which refer to donor atoms other than ones in the first row. It is seen that all the crosses are well above the reference curve, indicating greater reactivity than is predicted by the ΔG° values. The maximum deviation, about 7.5 kcal, is shown by I^-. These reactive nucleophiles are all soft, whereas all the first-row donors in Figure 1.2, except CN^-, are hard nucleophiles. The deviations shown by the crosses are examples of the rule "softness is more important in rates than in equilibria."

By comparing $\Delta G^\ddagger$ with ΔG°, we hope to have accounted for the dramatic effect of the solvent on lowering the reactivity of hard bases. Looking at Reaction (1.25), it appears that the enhanced reactivity of soft bases must be accounted for by favorable interactions between B and B′ in the TS, B′ : A : B. But this is just the symbiotic effect, since the leaving group, B, is iodide ion, which is soft. The existence of this symbiotic effect on rates has been known for some time.[42]

We can predict that the enhanced reactivity of soft bases will vanish if B is a hard base, such as F^-. The data for reaction in H_2O show just such an effect (Table 1.9).[43] The same large decreases occur when the leaving group, B, is sulfate, nitrate or sulfonate, all hard O-donors.

Table 1.9 Values of log *k* for Reaction of Bases in Water

	log k [$M^{-1}\,s^{-1}$]	
Base	CH_3F	CH_3I
H_2O	−10.8	−8.9
OH^-	−6.1	−4.3
Cl^-	−8.6	−5.5
I^-	−7.7	−3.5

The easiest way to identify symbiosis is to examine a ratio such as k_{CH_3I}/k_{CH_3F}, which will be large for soft nucleophiles and small for hard ones. A convenient probe is an ambident nucleophile which usually has a hard site and a soft site. An example is Me_2SO, which reacts at oxygen with methyl benzenesulfonate and at sulfur with methyl iodide. These changes in selectivity have been found in all solvents, including dipolar, aprotic ones.[44]

Such additional stabilization should appear in the gas-phase energy barrier. This is difficult to demonstrate because of the large effect of exothermicity, and because the stabilization is usually small. For example, in the gas-phase reactions of CH_3Cl with CH_3O^- and CH_3S^-, the barrier height is 2.6 kcal/mol higher for the soft CH_3S^-[40]. The reaction with CH_3O^- is 13 kcal/mol more exothermic, which could reduce its barrier by 5–6 kcal/mol, relative to CH_3S^-. The expected TS stabilization is only about 3 kcal/mol (see above), which is not inconsistent.

Symbiotic effects are a serious problem in the use of the Marcus equation to predict rates of methyl group transfers.[45] This otherwise useful equation predicts that for a pair of reactions such as

$$CH_3X + Y \rightarrow CH_3Y + X \qquad k_X \tag{1.32}$$

$$CH_3Z + Y \rightarrow CH_3Y + Z \qquad k_Z \tag{1.33}$$

that the ratio k_X/k_Z should be constant, or nearly so.[46] But we have just seen that the ratio k_I/k_F can change by a factor of 200 when Y changes fdrom OH^- to I^-. In other cases [47] the ratio changes by as much as 10^6. The theory behind the Marcus equation does not allow for interactions between entering and leaving groups.

OTHER APPLICATIONS

In one way or another, the HSAB Principle has found many applications in most areas of chemistry.[48] Usually these depend on the implicit use of Equation (1.7).

Unfortunately this equation is as quantitative as HSAB gets. There was no exact definition of hardness, η, and no operational definition which allowed it to be measured. The values in Tables 1.4 and 1.5 are useful only in ordering acids and bases. They are not transferable as numbers to any other examples. Therefore applications have always been qualitative. Even so, the HSAB concept has been extremely useful in correlating and understanding a great deal of experimental data.

Often this has been done by simply providing another perspective to view a body of information. Consider the case of the reactivity of metals, as given by the electromotive series:

$$\mathrm{K(s)} = \mathrm{K^+(aq)} + \mathrm{e^-} \qquad \Delta E^\circ = 2.93\,\mathrm{V} \tag{1.34}$$

$$\mathrm{Ag(s)} = \mathrm{Ag^+(aq)} + \mathrm{e^-} \qquad \Delta E^\circ = -0.80\,\mathrm{V} \tag{1.35}$$

What causes the large difference of 3.73 V, or 86 kcal/mol? Let us break the overall reaction into two parts, one requiring energy and the other giving back:

		K	Ag
$\mathrm{M(s)} = \mathrm{M^+(g)} + \mathrm{e^-}$	ΔH_1 [kcal/mol]:	120	243
$\mathrm{M^+(g)} = \mathrm{M^+(aq)}$	ΔH_2 [kcal/mol]:	79	116
		41	127

We see that potassium has an energy shortfall of 41 kcal/mol, but silver has a much larger one of 127 kcal/mol. The difference is equal to the 86 kcal/mol of free energy given by ΔE°, since entropy effects nearly cancel. Thus silver is less reactive than potassium partly because it is harder to ionize, due to greater electronegativity, but also because it does not interact with the solvent water as strongly as expected. Since water is a hard base and Ag^+ is a soft acid, such a result is reasonable.

Looking at the electronegative series of the elements, we see that the noble metals, Hg, Au, Pt, Ir, Pd, Os, Rh and Ru, all form metal ions which are classified as soft. Their heats of hydration are all less than expected and their inertness is thus explained.

The advantage of this viewpoint is that it focuses attention on the solvent. It suggests that by changing the solvent, we might invert the reactivities of the alkali metals, which form hard metal ions, and the Cu, Ag, Au triad. Even more easily, of course, by adding soft bases such as CN^- to water we can increase the reactivity of the noble metals.

The whole subject of solubility and solvation energies can be looked at in HSAB terms. The rule "*Similia similibus dissoluntur*" can be replaced by "Hard solvents dissolve hard solutes, and soft solvents dissolve soft solutes". This requires examination of the generalized acid–base properties of solvents. Most solvents have a donor site and an acceptor site, which may differ widely in both hardness and strength.

Hydrogen bonding is an acid–base phenomenon. We find that the strongest hydrogen bonds are formed between hard acids and hard bases, such as H_2O and OH^-. The smaller heat of hydration of SH^-, although explained by simple size effects, is also consistent with the HSAB viewpoint. The conclusion is that hydrogen bonding is mainly an electrostatic interaction. This is in contrast to the Mulliken approach, which stresses charge transfer.

Empirical softness values have been assigned to some 90 common solvents.[48] The scale is defined as the difference in the free energy of transfer of a hard cation (Na^+ or K^+) from water to a given solvent and the corresponding quantity for a soft cation (Ag^+). The scale can be used in linear solvation energy correlations, or simply to estimate solubilities.

Stabilization of certain oxidation states for an element can be predicted by HSAB. For example, to have a Ni(IV) compound, we will need hard ligands, or counter ions, such as F^- or O^{2-}. To have Ni(0), we need soft ligands, such as CO, PR_3, AsR_3 or R_2S. The growth of organometallic chemistry of the transition metals, which was just under way in 1963, was facilitated by realizing that ligands such as CH_3^-, H^- and C_2H_4 were all soft, and that they would need soft metal centers to which to bind. The metal could be made soft by a low oxidation state, which in turn could be stabilized by other soft ligands.

An unusual property of transition metal hydrides is that they are often quite strong Brönsted acids.[49] A typical example would be

$$H_2Fe(II)(CO)_4 = H^+ + HFe(0)(CO)_4^- \qquad pK_a = 4.4 \qquad (1.36)$$

The ionization reaction is also a typical reductive elimination, since the formal oxidation state of the metal decreases by two units. We can predict that hard ligands will stabilize Fe(II) in preference to Fe(0), and will *reduce* the acidity of $H_2Fe(CO)_6$.

This kind of prediction has been amply verified for other cases.[49] Consider the rhodium(III) complexes

$$Rh(NH_3)_4(H_2O)H^{2+} \qquad pK_a > 14$$

$$Rh(bipy)_2(H_2O)H^{2+} \qquad pK_a = 9.5$$

$$Rh(CNR)_4(H_2O)H^{2+} \qquad pK_a < 0$$

The effect is quite large, even though the net charges are constant. The order of increasing η is isocyanide < bipyridine (bipy) < ammonia.

Organometallic compounds of the transition metals are involved in many important examples of homogeneous catalysis, such as the Wacker process and the oligomerization of olefins. These require that an organic molecule, such as ethylene, be coordinated to a metal atom holding another potential reactant.

A good strategy is to have a hard ligand attached to a soft metal atom. The hard ligand is often a solvent molecule such as H_2O or CH_3CN. Since these will be labile, vacant coordination sites for the organic reactant are readily available.[50] Soft ligands, present in the solution, will act as catalyst poisons.

Saville has developed rules for chemical reactions that can also be used to select homogeneous catalysts.[51] Consider the four-center reaction

$$B' + A:B + A' = B':A + B:A' \tag{1.37}$$

Reaction is facilitated if A : B is a mismatched pair in terms of hardness, and if the hardness of B′ matches that of A and that of A′ matches B. An example would be the cleavage of ethers by hydrogen iodide:

$$I^- + C_2H_5OC_2H_5 + H^+ \rightleftharpoons C_2H_5I + C_2H_5OH \tag{1.38}$$

The ready hydrolysis of alkyl iodides makes this a catalytic reaction.

Heterogeneous catalysis is made much more understandable by considering the interactions between the surface atoms of the catalyst and the adsorbed reactants as acid–base reactions. The two main classes of catalysts are typified by the transition metals and by the acid clays. The bulk metals have atoms in the zero-valent state and are all soft acids. They are also soft bases, since they can donate electrons easily. Catalysts such as Al_2O_3–SiO_2 mixtures contain hard metal ions as acids, and hard oxide, or hydroxide, ions as bases.

Because of strong adsorption, we expect compounds of P, As, Sb, S and Se to be poisons for transition metal catalysts. Soft acids such as Hg^{2+} and Pb^{2+} will also be poisons. But poisons for the acid clays will be hard metal ions, and hard bases, such as NH_3, CO_3^{2-} and SO_4^{2-}.

Differences in reaction products, such as

$$C_2H_5OH \xrightarrow[\text{metal}]{\Delta} CH_3CHO + H_2 \tag{1.39}$$

$$C_2H_5OH \xrightarrow[\text{clay}]{\Delta} C_2H_4 + H_2O \tag{1.40}$$

are readily accounted for. The metal surfaces remove hydride ions (and return electrons to the substrate); the acid clay removes H^+ and OH^-.

Pauling's bond energy equation is one of the best-known empirical equations in chemistry:

$$D^\circ_{AB} = \tfrac{1}{2}(D^\circ_{AA} + D^\circ_{BB}) + 23(\chi_A - \chi_B)^2 \tag{1.41}$$

Pauling used it to assign electronegativity values, χ_A and χ_B, to the elements.[52] It still often used in various ways, in spite of many publications pointing out its inaccuracies.

Applied to the exchange reaction (Equation (1.16)), the Pauling equations gives

$$\Delta H = 46(\chi_{A'} - \chi_A)(\chi_B - \chi_{B'})\,\text{kcal/mol} \tag{1.42}$$

That is, the reaction will be exothermic if the products contain the least electronegative (EN) element combined with the most EN element. But this usually is the opposite of what Equation (1.7) will predict. Take the example

$$\text{LiI(g)} + \text{CsF(g)} = \text{CsI(g)} + \text{LiF(g)} \qquad \Delta H = -17\,\text{kcal/mol} \tag{1.43}$$

Equation (1.42) gives $\Delta H = 46(0.7 - 1.0)(2.5 - 4.0) = +21$ kcal/mol.

Equation (1.7) will correctly predict that Reaction (1.43) is exothermic, assuming that Cs^+ is softer than Li^+, just as I^- is softer than F^-, and as shown in Table 1.4. Many similar examples can be cited.[52] If Equation (1.41) is used along with Pauling's EN values to calculate Δ in Table 1.4, there will be errors as large as 80 kcal/mol.

The limited validity of Equation (1.41) is for bonds where $(\chi_A - \chi_B)$ is small, that is, for covalent bonds. It is useless for bonds that are highly ionic, as we have in Reaction (1.43).[54] There are improved versions of Pauling's equations available, which are much better for ionic bonds.[55] In spite of these criticisms, the EN values calculated by Pauling using Equation (1.41) are perfectly reasonable and are accepted as the standard. This is a tribute to Pauling's insight (and to a careful selection of data to be used!).

SUMMARY

In retrospect it seems clear that the original concept of hardness and softness is an example of "fuzzy logic." In spite of its name, fuzzy logic is a respected branch of mathematics, which was essentially invented in 1965 by L. A. Zadeh. It is a method of reasoning and making decisions when the available information is not precise enough to use numbers, or definite statement such as "Yes" or "No." On the other hand there is some information which is qualitative in nature and insufficient to make definite statements.

Classical mathematics requires that a concept has a precise definition which partitions a class of objects into two classes: (a) those that belong, and (b) those that do not belong. But the real world is filled with concepts such as many, old, slow, large. These do not have sharply defined boundaries, but still convey information. Fuzzy logic deals with the application of such knowledge.

As might be expected, there was at first considerable criticism and rejection of Zadeh's proposals. Today, however, there are many papers being published that concern the development and application of fuzzy logic. There are a number of important uses already well established, particularly in technology, manufacturing and finance. Applications in science will surely follow. It is of interest to note the recent appearance of a book entitled *Fuzzy Logic in Chemistry*.[56]

Nevertheless, it is a great advantage in science to have quantitative definitions so that one can measure what one is speaking about, and express it in numbers. Fortunately this is what has happened to chemical hardness. The means by which this has come about lies in density functional theory. This will be the topic of the next chapter.

REFERENCES

1. R.G. Pearson, *J. Am. Chem. Soc.*, **85**, 3533 (1963).
2. S. Ahrland, J. Chatt and N.R. Davies, *Quart. Rev.* (London), **12**, 255 (1958).
3. G. Schwarzenbach, *Adv. Inorg. Che. Radiochem. 3*, **1** (1961).
4. J.O. Edwards, *J. Am. Chem. Soc.*, **76**, 1540 (1954).
5. R.G. Pearson, *Chemistry in Britain*, **3**, 103 (1967).
6. R.S. Drago and B.B. Wayland, *J. Am. Chem. Soc.*, **87**, 3571 (1965).
7. G. Klopman, *J. Am. Chem. Soc.*, **90**, 223 (1968).
8. R.G. Pearson, *J. Chem. Ed.*, **45**, 643 (1968).
9. R.G. Pearson and R.J. Mawby, *Halogen Chemistry*, V. Gutmann, Ed., Academic Press, New York, 1967, Vol. 3, p. 61ff.
10. R.G. Pearson, *J. Am. Chem. Soc.*, **110**, 7684 (1988).
11. A.F. Bochkov, *Zhur. Org. Khim.*, **22**, 2041 (1986). A.S. Peregrudov et al., *J. Organomet. Chem.*, **471**, C1 (1994).
12. R.R. Squires, *J. Am. Chem. Soc.*, **107**, 4385 (1985); P.B. Armentrout, L.F. Halle and J.L. Beauchamp, ibid., **103**, 6501 (1983).
13. R.G. Pearson, *J. Mol. Struct. (Theochem.)*, **300**, 519 (1993).
14. P.K. Chattaraj and P.v.R. Schleyer, *J. Am. Chem. Soc.*, **116**, 1067 (1994).
15. S.G. Lias, J.F. Liebman and R.D. Levin, *J. Phys. Chem. Ref. Data*, **13**, 695 (1984).
16. R.W. Taft, J.F. Wolf, J.T. Beauchamp, G. Scorrano and E.M. Arnett, *J. Am. Chem. Soc.*, **100**, 1240 (1978).
17. T.B. McMahon, T. Heinis, G. Nicol, J.K. Hovey and P. Kebarle, *J. Am. Chem. Soc.*, **110**, 7591 (1988); R.G. Pearson, ibid., 7684; C.A. Deakyne, and M. Meot-Ner, *J. Phys. Chem.*, **94**, 232 (1990).
18. S.J. Ashcroft and C.T. Mortimer, *Thermochemistry of Transition Metal Complexes*, Academic Press, New York, 1970; J.J. Christensen, D.J. Eatough and R.W. Izatt, *Handbook of Metal Ligand Heats*, Marcel Dekker, New York, 1978.
19. S. Ahrland, *Helv. Chim. Acta*, **50**, 306 (1963).
20. P. Gerding, *Acta Chem. Scand.*, **20**, 2771 (1966); R. Hancock and R. Marsicano, *J. Chem. Soc., Dalton Trans.*, **1832** (1976).
21. H. Irving and R.J.P. Williams, *Nature* (London), **162**, 146 (1948).
22. R.J.P. Williams, *Discuss. Faraday Soc.*, **26**, 123 (1958).
23. R.P. Hancock and A.E. Martell, *Adv. Inorg. Chem.*, **42**, 89 (1995).
24. R.D. Hancock and F. Marsicano, *Inorg. Chem.*, **17**, 560 (1978); idem, ibid. **19**, 2709 (1980).

25. V. Jonas, G. Frencking and M. T. Reetz, *J. Am. Chem. Soc.*, **116**, 8741 (1994).
26. C.K. Jorgensen, *Inorg. Chem.*, **3**, 1201 (1964).
27. J.M. Pratt and R.G. Thorp, *J. Chem. Soc. A*, 187 (1966).
28. J. Hine, *J. Am. Chem. Soc.*, **85**, 3239 (1963).
29. A.E. Reed and P.v.R. Schleyer, *J. Am. Chem. Soc.*, **109**, 7362 (1987).
30. H.B. Schlegel, *J. Phys. Chem.*, **88**, 6254 (1984).
31. R.G. Pearson, *Inorg. Chem.*, **12**, 712 (1972).
32. F. Basolo and R.G. Pearson, *Mechanisms of Inorganic Reactions*, 2nd edn., John Wiley, New York, 1967, Chapter 5.
33. J. Chatt and B.T. Heaton, *J. Chem. Soc. A*, 2745 (1968).
34. J.O. Edwards and R.G. Pearson, *J. Am. Chem. Soc.*, **84**, 16 (1962).
35. W.P. Jencks and J. Carriuolo, *J. Am. Chem. Soc.*, **82**, 1778 (1960).
36. R.G. Pearson, H. Sobel and J. Songstad, *J. Am. Chem. Soc.*, **90**, 319 (1968).
37. R.G. Pearson, *J. Org. Chem.*, **52**, 2131 (1987).
38. J.E. Leffler and E. Grunwald, *Rates and Equilibria of Organic Reactions*, John Wiley, New York, 1963.
39. For references see S. Wolfe, D.J. Mitchell, H.B. Schlegel, *J. Am. Chem. Soc.*, **103**, 7692, 7694 (1981).
40. M. Pellerite and J.I. Brauman, *J. Am. Chem. Soc.*, **102**, 5993 (1980); ibid., **105**, 2672 (1983).
41. J. Chandrasekhar, S.F. Smith and W.L. Jorgensen, *J. Am. Chem. Soc.*, **108**, 154 (1983).
42. R.G. Pearson and J. Songstad, *J. Org. Chem.*, **32**, 2899 (1967).
43. J. Koskikallio, *Acta Chem. Scand.*, **26**, 1201 (1972).
44. L.H. Sugemyr and J. Songstad, *Acta Chem. Scand.*, **26**, 4179 (1972); A.J. Parker, *Chem. Rev.*, **69**, 1 (1969).
45. For a review see W.J. Albery, *Annu. Rev. Phys. Chem.*, **31**, 227 (1980).
46. E.S. Lewis, M.L. McLaughlin and T.A. Douglas, *J. Am. Chem. Soc.*, **107**, 6668 (1985).
47. R.G. Pearson and P.E. Figdore, *J. Am. Chem. Soc.*, **102**, 1541 (1980).
48. Y. Marcus, *J. Phys. Chem.*, **91**, 4422 (1987).
49. R.G. Pearson, *Chem. Rev.*, **85**, 41 (1985); idem, *Bonding Energetics in Organometallic Compounds*, T. Marks, Ed., ACS Symposium Series 428, American Chemical Society, Washington, DC, pp. 260–261.
50. W.J.A. Davies and F.R. Hartley, *Chem. Rev.*, **81**, 79 (1981); S.L. Randall *et al.*, *Organometallics*, **13**, 5088 (1994).
51. B. Saville, *Angew. Chem., Int. Ed. Engl.*, **6**, 928 (1967).
52. L. Pauling, *The Nature of the Chemical Bond*, 3rd Edn., Cornell University Press, Ithaca, 1960, pp. 88–105.
53. R.G. Pearson, *Chem. Commun.*, **65** (1968).
54. R.S. Drago, N. Wong and D.C. Ferris, *J. Am. Chem. Soc.*, **113**, 1970 (1991).
55. R.L. Matcha, *J. Am. Chem. Soc.*, **105**, 4859 (1983). D.W. Smith, *J. Chem. Ed.*, **67**, 911 (1990).
56. D.H. Rouvray (Ed.), *Fuzzy Logic in Chemistry*, Academic Press, New York, 1997.

2 Density Functional Theory

INTRODUCTION

Density functional theory (DFT) is a form of quantum mechanics which uses the one-electron density function, ρ, instead of the more usual wave function, ψ, to describe a chemical system.[1] Such a system is any collection of nuclei and electrons. It may be an atom, a molecule, an ion, a radical or several molecules in a state of interaction.

Hohenberg and Kohn proved in 1964 that the ground-state energy of a chemical system is a functional of ρ only.[2] A functional is a recipe for turning a function into a number, just as a function is a recipe for turning a variable into a number. For example, the energy is also a functional of the wave function. The variational method is one recipe for turning ψ into a number, E.

$$E = \frac{\langle \psi \hat{H} \psi \rangle}{\langle \psi \psi \rangle} \tag{2.1}$$

$\hat{H}$ is the many-electron Hamiltonian operator, just as ψ is the many-electron wave function. The angle-brackets mean integration over the electronic coordinates.

The density, ρ, can be obtained by squaring ψ and integrating over the coordinates of all the electrons but one. This is then multiplied by N, the total number of electrons, to get the number of electrons per unit volume, ρ, which is a function of the three space coordinates only. It is a quantity easily visualized and experimentally measurable by X-ray diffraction, though the accuracy is not adequate for chemical purposes.

The basic notion that the energy is expressible in terms of the density goes back to the Thomas–Fermi atom in 1927–1928.[3] In this model the kinetic energy and the potential energy are expressed in terms of ρ. There is a definite recipe for obtaining the total energy and other properties of the atom. The first density-based scheme to be used for more than one atom is the Hartree–Fock–Slater, or $X\alpha$, method.[4] Slater proposed that the effects leading to electron exchange energies and correlation energies be given by a function proportional to $\rho^{1/3}$.

This approximation arose because of the need to simplify the quantum mechanics of the solid state. DFT has been the almost exclusive method used in solid-state physics since the 1950s. The reason is that the wave function depends on $3N$ space coordinates, as well as spin coordinates, while ρ depends only on three space coordinates. A similar advantage exists whenever N is large, as in large molecules.

The advantage is great when calculations are made with large digital computers. The time required scales as N^4 using Hartree–Fock (HF) wave functions, and N^3 using electron density. As a result, DFT has emerged as an alternative *ab-initio* method to HF-based variational methods. Another advantage is that relativistic corrections are easily made in DFT, whereas they are very difficult in HF.[5] Such corrections are very important for the transition metals, and other heavy atoms.

The accuracy obtained in all cases depends on the details of the method used. The most accurate calculations are those obtained by HF methods with full correlation energy corrections. (The correlation energy is defined as the difference between the HF energy and the exact energy.) But these are only practical for very small values of N, since computer times now scale as N^5. The best DFT methods available at present are equal to the best practical HF-based methods available; that is, there is some correlation energy included.[6] At the same time the computer time required is 10 to 100 times less for DFT calculations, depending on N. It is hard to avoid the conclusion that density functional theory will almost completely replace wave function theory in the area of *ab-initio* calculations on molecules.

In terms of ultimate accuracy, or exactness, of the theories, Hohenberg and Kohn showed that DFT was an exact theory in the same sense as wave theory.[2] The Schrödinger equation

$$\hat{H}\psi_0 = E_0\psi_0 \tag{2.2}$$

reveals that all properties of the ground state are functions only of N and $v(r)$, the potential due to the nuclei. But it can be proved that $\rho(r)$ determines both N and $v(r)$.[2] Thus $\rho(r)$ also determines ψ_0 and all ground-state properties. It can also be proved that a trial electron density, $\tilde{\rho}(r)$, which is not the exact $\rho(r)$, will give an energy higher than the exact energy, E_0.

We can write the energy as

$$E(\rho) = V_{ne}(\rho) + J(\rho) + T(\rho) + E_{xc}(\rho) \tag{2.3}$$

where all terms are explicit functionals of the electron density. V_{ne} is the nuclear-electron potential energy, J is the classical part of the electron–electron repulsion energy, E_{xc} is the so-called exchange-correlation energy and T is the kinetic energy. Given the function ρ, we can readily calculate V_{ne} and J, but we do not know the exact functional dependence of T and E_{xc} on ρ. Hence we cannot calculate the energy without approximations.

The usual method of proceeding is to solve for the Kohn–Sham orbitals, u_i:[7]

$$\hat{h}_{KS}u_i = \varepsilon_i u_i \tag{2.4}$$

The one-electron Hamiltonian is given by

$$\hat{h}_{KS} = v(r) + v_e(r) + v_{xc}(r) - \tfrac{1}{2}\nabla^2 \tag{2.5}$$

where $v_e(r)$ is the classical potential due to the electrons and v_{xc} is the potential leading to E_{xc} in Equation (2.3). Fortunately very good approximations are already available for v_{xc}.[3] The kinetic energy is given by $-\frac{1}{2}\nabla^2$, in atomic units.

Since we now have a one-electron problem, the Kohn–Sham equations (2.4) can be solved in a self-consistent manner. We obtain a set of orbitals and their energies, much as in HF theory. The density function, $\rho(r)$, can be found as the sum of the squares of the u_i, for the occupied orbitals. From $\rho(r)$ the expectation value of the energy can be found, as well as other one-electron properties. Just as in the HF method, the total electronic energy is equal to the sum of the energies of the occupied orbitals, minus a correction because the electron–electron interactions have been counted twice.

Strictly speaking, the Kohn–Sham (KS) orbitals are fictitious entities, created by a certain mathematical procedure. Of course, the same can be said about molecular orbitals (MOs) in HF theory. But we know that MOs are given reality by their successful use in many applications. The KS orbitals and the HF MOs are not the same, but there is a one-to-one correspondence, and their orbital energies are similar. It would appear that KS orbitals will one day be used in the same way as MOs are now.

Since in principle DFT theory is exact and the KS equations are exact, if the exact v_{xc} is used in Equation (2.5) we can expect that the KS orbitals will eventually prove superior for some applications. At least in one case, we know that this is true. The orbital energy of the highest occupied KS orbital will approach the negative of the first ionization potential exactly.[9]

The foregoing has emphasized the value of DFT in the accurate calculation of molecular properties. But there is another aspect which has not been mentioned. DFT is rich in conceptual content. Many of the basic concepts of chemistry, including hardness and electronegativity (EN), appear simply and naturally. Thus DFT is useful not only for calculations but also for understanding them. Much of this is due to the pioneering work of R.G. Parr and his collaborators.

DFT AND THE CONCEPTS OF CHEMISTRY

Consider a chemical system consisting of several nuclei and N electrons. The nuclei generate a potential v. Holding the nuclei fixed in position, the ground-state electron density function ρ is that which satisfies the variational equation

$$\delta[E[\rho] - \mu N[\rho]] = 0 \tag{2.6}$$

The quantity μ is the Lagrange multiplier that ensures that the integral of ρ over the volume is equal to N. It follows that

$$\mu = \left|\frac{\delta E[\rho]}{\delta\rho}\right|_v = v(r) + \left|\frac{\delta F[\rho]}{\delta\rho}\right|_v \tag{2.7}$$

where $F[\rho]$ is the sum of the last three terms in Equation (2.3). The use of δ in Equation (2.7) implies that the terms are to be read as "the functional dependence of $E[\rho]$ and $F[\rho]$ on ρ". We can also write the total differential of E as

$$\mathrm{d}E = \mu \mathrm{d}N + \langle \rho \mathrm{d}v \rangle \tag{2.8}$$

since E is a function of N and v only. Accordingly we have the new relationship

$$\mu = \left(\frac{\partial E}{\partial N}\right)_v = -\chi \tag{2.9}$$

The quantity μ is called the electronic chemical potential. The name comes from the thermodynamic equation

$$T\mathrm{d}S = \mathrm{d}E + P\mathrm{d}V - \mu_T \mathrm{d}N \tag{2.10}$$

At zero pressure and temperature, we also have $\mu_T = (\partial E/\partial N)$. In this case N is the number of molecules in the system, and μ_T is the ordinary chemical potential of thermodynamics. The electronic chemical potential of a single molecule plays somewhat the same role. At equilibrium μ must be constant everywhere, and ρ will be the correct electron density for the ground state. The quantity χ is called the absolute electronegativity, for reasons that will become clear.[8]

The definition of μ in Equation (2.9) is much preferable to that in Equation (2.7), which gives μ in terms of the functional dependence of $E(\rho)$ on ρ, which is not known. But we do know something about the variation of E with N. Figure 2.1 shows a plot of the energy of a chemical system as a function of the number of electrons. The energies are all negative, with zero energy way up on top. Experimentally we only know points on the curve for integral values of N, from data such as ionization potentials, I, and electron affinities, A. If we connect these points by a smooth curve, then $(\partial E/\partial N)$ is simply the instantaneous slope of the curve.

We do not know this instantaneous slope, nor is it straightforward to calculate it, since the Schrödinger equation is defined only for integral values of N. However, if we pick a starting point, such as the neutral species, we know the mean slope from N to $(N-1)$ electrons. It is the negative of the ionization

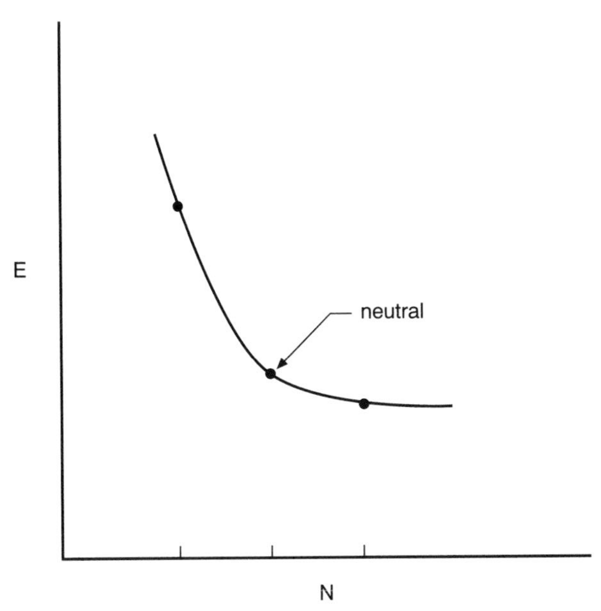

Figure 2.1 Plot of the total electronic energy vs. the number of electrons for a fixed collection of nuclei.

potential, I. Also the mean slope from N to $(N+1)$ is the negative of the electron affinity, A. By the method of finite differences we can estimate the slope at N as

$$-\mu = -(\partial E/\partial N)_v \simeq (I + A)/2 = \chi_M \tag{2.11}$$

But this is simply the Mulliken EN, χ_M.[10] In addition, there is a previous history of calling $(\partial E/\partial N)$ the electronegativity.

Actually, if we define the absolute EN, χ, as $(I + A)/2$, it is not quite the same as Mulliken's EN. He was interested in a scale which could be used to estimate the polarity of chemical bonds, just as Pauling was. Mulliken's I and A are for supposed valence states of an atom or radical, such as it might have in a molecule. The DFT values of I and A are for the ground state of any system, atom, ion, radical or molecule. Also, since v is to be constant, they refer to vertical values and not adiabatic ones.

Besides Equation (2.11), there is a second reason to call χ the absolute EN. If two systems, C and D, approach each other, with different values, μ_C^0 and μ_D^0, for their electronic chemical potentials, there must be a flow of electrons from one to the other until the potentials are equalized, i.e. $\mu_C = \mu_D$. This is the condition for equilibrium. But this means that the ENs must also equalize: $\chi_C = \chi_D$. Thus we have a proof of EN equalization, originally proposed by Sanderson as a postulate.[12]

To make use of this principle of equalization, we must know how μ changes as we change the number of electrons in the subsystems, C and D. But that is just what Figure 2.1 tells us. The slope is not constant, but becomes smaller (less negative) as N increases. Equation (2.12) gives the curvature of the plot of E versus N:

$$(\partial\eta/\partial N)_\nu = (\partial^2 E/\partial N^2)_\nu = 2\eta \tag{2.12}$$

The quantity η is called the absolute, or chemical, hardness.[13] From the method of finite differences we obtain the operational definition of η:

$$\eta \simeq (I - A)/2 \tag{2.13}$$

The units of μ, η and χ are all the same, usually eV.

For isolated reactants, C and D, we can now write

$$\mu_C = \mu_C^0 + 2\eta_C \Delta N \tag{2.14}$$

$$\mu_D = \mu_D^0 - 2\eta_D \Delta N \tag{2.15}$$

where ΔN is the fractional number of electrons transferred from D to C. Applying electronegativity equalization, or $\mu_C = \mu_D$, we find that

$$\Delta N = \frac{(\chi_C^0 - \chi_D^0)}{2(\eta_C + \eta_D)} \tag{2.16}$$

Thus the difference in electronegativity drives the process, and the sum of the hardness parameters acts as a resistance. Equation (2.16) is a chemical form of Ohm's Law. It shows hat electrons will flow from the system of low EN to that of high EN, as expected. It also shows that Equation (2.12) is a reasonable definition of hardness, since the common meaning of hardness is resistance to deformation or change.

The most immediate reason for Equation (2.12), however, was that it agreed with the chemical observations that led to the concept of hard and soft acids and bases. For example, consider Ca^{2+} and Fe^{2+}. The third ionization potential, I_3, and the second, I_2, would be I and A, respectively (Table 2.1). Accordingly, Fe^{2+} is much softer than Ca^{2+}, as expected. Also Ca^{2+} is more EN than Fe^{2+}, meaning that it is much less likely to find Ca^{3+} than Fe^{3+}. Similar results are found or almost all the metal ions.

Fortunately for the further development of hardness and DFT, a large number of I and A values have become available. It turns out that most common molecules have negative values for their electron affinity; that is, energy is required to force an electron on to the molecule. The technique of electron

Table 2.1 Ionization Potentials, χ and η for Ca^{2+} and Fe^{2+}

	Ca^{2+}	Fe^{2+}
I_2 [eV]	11.87	16.18
I_3 [eV]	51.21	30.64
$\chi^{(a)}$ [eV]	31.54	23.41
$\eta^{(b)}$ [eV]	19.67	7.23

(a) $\chi = (I_2 + I_3)/2$ (Equation (2.11)).
(b) $\eta = (I_3 - I_2)/2$ (Equation (2.13)).

transmission spectroscopy has recently been developed to measure negative A values.[14] These are always vertical values, as required by DFT. Positive A values are almost always adiabatic results, which can differ appreciably from vertical ones. Ionization potentials, which are always positive, are usually the adiabatic values, but it is sometimes not clear whether reported numbers are adiabatic or vertical.

In spite of these difficulties, Table 2.2 presents χ and η data for a number of neutral molecules. They are arranged in order of decreasing χ, so that Lewis acids are at the top and Lewis bases at the bottom. The order shown is not to be taken as an order of acid or base strength, but of preference for accepting electrons over giving them up. Some molecules, such as SF_6, are very inert. But when SF_6 does react, it reacts with bases. Again, the hardness numbers are as expected from chemical evidence. H_2O is harder than H_2S; NH_3 is softer than H_2O but harder than PH_3; and so on.

There are serious problems in ranking ions, both anions and cations, in a way which is commensurate with Table 2.2. First, the data are usually unavailable. One cannot measure the electron affinity of Br^-, since Br^{2-} does not exist. Except for monatomic cations, second ionization potentials usually cannot be measured. But even more seriously, the finite difference method used in Equations 2.11 and 2.12 weights the gain and loss of an electron equally. This is inappropriate for ions, where a strong bias for unidirectional flow of charge exists.

It is tempting to assume that $\chi = I$ for anions and $\chi = A$ for cations, and probably more correct than $(I + A)/2$. But then it becomes difficult to assign a number to the hardness. Also, such χ values could not be used in comparing ions with neutral species, as in Equation (2.16). Note that the results given above for Ca^{2+} and Fe^{2+} depend on I_3, which is for removal of an electron which is not in the valence shell for Ca^{2+}. Such a number must have limited chemical significance.

Looking at several examples, as in Table 2.3, makes the situation clearer. Both χ and η tell us that it is easier to change the oxidation state of Cu^+, Fe^{2+} and Tl^{3+}, than those of Na^+, Ca^{2+} and Al^{3+}. The increasing values of χ as the positive charges increase are expected. In the case of Tl^{3+}, χ is so large that only

Table 2.2 Experimental Parameters for Molecules [eV]

Molecule	$I^{(a)}$	$A^{(a)}$	χ	η
F_2	15.70	3.1	9.4	6.3
SF_6	15.4	0.5	8.0	7.4
O_3	12.4	2.1	7.25	5.2
SO_3	12.7	1.7	7.2	5.5
Cl_2	11.6	2.4	7.0	4.6
H_2	15.4	−2.0	6.7	8.7
SO_2	12.3	1.1	6.7	5.6
N_2	15.58	−2.2	6.70	8.9
Br_2	10.56	2.6	6.6	4.0
C_2N_2	13.37	−0.58	6.40	6.98
O_2	12.07	0.44	6.25	5.82
BF_3	15.8	−3.5	6.2	9.7
CO	14.0	−1.8	6.1	7.9
CS	11.7	0.2	6.0	5.8
I_2	9.4	2.6	6.0	3.4
BCl_3	11.60	0.33	5.97	5.64
HNO_3	11.03	0.57	5.80	5.23
CH_3NO_2	11.13	0.45	5.79	5.34
PF_3	12.3	−1.0	5.7	6.7
HCN	13.6	−2.3	5.7	8.0
BBr_3	10.51	0.82	5.67	4.85
PBr_3	9.9	1.6	5.6	4.2
S_2	9.36	1.66	5.51	3.85
$C_6H_5NO_2$	9.9	1.1	5.5	4.4
$CHCl_3$	11.4	−0.3	5.5	5.9
CCl_4	11.47	−0.49[b]	5.50	6.00
PCl_3	10.2	0.8	5.5	4.7
N_2O	12.9	−2.2	5.4	7.6
Acrylonitrile	10.91	−0.21	5.35	5.56
CS_2	10.08	0.62	5.35	5.56
CH_2	10.0	0.6	5.3[e]	4.7[e]
HI	10.5	0.0	5.3	5.3
CH_2Cl_2	11.3	−1.1	5.1	6.2
CO_2	13.8	−3.8	5.0	8.8
HF	16.0	−6.0	5.0	11.0
HCHO	10.9	−0.9	5.0	5.9
CH_2S	9.3	0.5[d]	4.9	4.4
CH_3I	9.5	0.2	4.9	4.7
CH_3Br	10.6	~ −1.0	4.8	5.8
SiH_4	11.7	−2.0	4.8	6.8
HCl	12.7	−3.3	4.7	8.0
CH_3CN	12.2	−2.8	4.7	7.5
HCO_2CH_3	11.0	−1.8	4.6	6.4
CH_3CHO	10.2	−1.2	4.5	5.7
C_2H_4	10.5	−1.8	4.4	6.2

Table 2.2 (continued)

Molecule	$I^{(a)}$	$A^{(a)}$	χ	η
Butadiene	9.1	−0.6	4.3	4.9
H_2S	10.5	−2.1	4.2	6.2
C_2H_2	11.4	−2.6	4.4	7.0
$HCONH_2$	10.2	−2.0	4.1	6.1
Styrene	8.47	−0.25	4.11	4.36
CH_3COCH_3	9.7	−1.5	4.1	5.6
PH_3	10.0	−1.9	4.1	6.0
C_6H_6	9.3	−1.2	4.1	5.3
AsH_3	10.0	−2.1[d]	4.0	6.1
c-C_3H_6	10.5[f]	−2.6[c]	4.0	6.6
Toluene	8.8	−1.1	3.9	5.0
CH_3Cl	11.2	−3.7	3.8	7.5
p-Xylene	8.4	−1.1	3.7	4.8
1,3,5-Trimethyl-benzene	8.40	−1.03	3.69	4.72
Cyclohexene	8.9	−2.1	3.4	5.5
DMF	9.1	−2.4	3.4	5.8
CH_3F	12.5	−6.2	3.2	9.4
H_2O	12.6	−6.4	3.1	9.5
$(CH_3)_3As$	8.7	−2.7	3.0	5.7
$(CH_3)_3P$	8.6	−3.1	2.8	5.9
$(CH_3)_2S$	8.7	−3.3	2.7	6.0
NH_3	10.7	−5.6	2.6	8.2
CH_4	12.7	−7.8	2.5	10.3
CH_3OH	10.9	−6.2	2.3	8.5
$C(CH_3)_4$	10.4	−6.1	2.2	8.3
$(CH_3)_2O$	10.0	−6.0	2.0	8.0
CH_3NH_2	9.0	−5.3[d]	1.9	7.2
$(CH_3)_3N$	7.8	−4.8	1.5	6.3

[a] Data from References 49 and 50, except as indicated.
[b] G.L. Gutsev and T. Ziegler, Can. J. Chem., *69*, 993 (1991).
[c] M. Allan, J. Am. Chem. Soc., *115*, 6418 (1993).
[d] S. Moran and G.B. Ellison, Int. J. Mass Spectrom., *80*, 83 (1988).
[e] Singlet state.

Table 2.3 Values of χ and η for Various Monatomic Cations

	Na^+	Cu^+	Ca^{2+}	Fe^{2+}	Al^{3+}	Tl^{3+}
χ [eV]	26.21	14.01	31.39	23.42	74.22	40.3
η [eV]	21.08	6.28	19.52	7.24	45.77	10.4

the *gain* of electrons can be expected. In short, there is considerable qualitative information given, even though χ and η cannot be used in Equation (2.16).

In general, it is not possible to add or remove a second electron from a small molecule to yield a stable species. In the case of a double positive charge, M^{2+}, the molecule dissociates into two singly charged ions. M^{2-} would simply not be formed in detectable amounts. In both cases, however, as M becomes very large, stable, multiply charged species can exist.

CORRELATION WITH MOLECULAR ORBITAL THEORY

There are still more reasons to believe that η, as defined in Equation (2.12), is indeed what is meant by chemical hardness. To understand this, it is necessary to see whether the chemical concepts derived by DFT are compatible with molecular orbital (MO) theory.[15] This theory is certainly the most widely used by chemists and is very successful in many areas. It is almost universally applied to explain structure and bonding, visible-UV spectra, chemical reactivity and detailed mechanisms of chemical reactions.

The best way to combine DFT and MO theory is to incorporate χ and η into the commonly used orbital energy diagrams. Figure 2.2(a) shows such a diagram. A typical case of a molecule where $I = 10\,\text{eV}$ and $A = -2\,\text{eV}$ is taken. Within the validity of Koopmans' theorem,[16] the frontier orbital energies are given by

$$-\varepsilon_{\text{HOMO}} = I \qquad \text{and} \qquad -\varepsilon_{\text{LUMO}} = A \tag{2.17}$$

The value of $\chi = 4\,\text{eV}$ is shown with changed sign as a dashed horizontal line in Figure 2.2(a). It falls exactly at the energy midpoint between the HOMO and the LUMO. The value of $\eta = 6\,\text{eV}$ is shown as a vertical dashed line. The energy gap between the HOMO and the LUMO is equal to 2η.

The above refers to a system where the HOMO is filled. Many radicals where the frontier orbital (SOMO) is half-filled, are somewhat different if I and A both refer to the SOMO. Figure 2.2(b) shows the case of a radical where $I = 10\,\text{eV}$, and $A = +2\,\text{eV}$. The SOMO orbital energy ($-10\,\text{eV}$), $\chi = 6\,\text{eV}$, and $\eta = 4\,\text{eV}$, are shown on the figure. The unknown energy of the LUMO plays no role. The quantity $(I - A)$ is just the mean inter-electronic repulsion of two electrons in the SOMO.

Apart from the radical cases, it would appear that Figure 2.2 offers a most graphic and concise way of defining what is meant by chemical hardness:

> Hard molecules have a large HOMO–LUMO gap, and soft molecules have a small HOMO–LUMO gap.

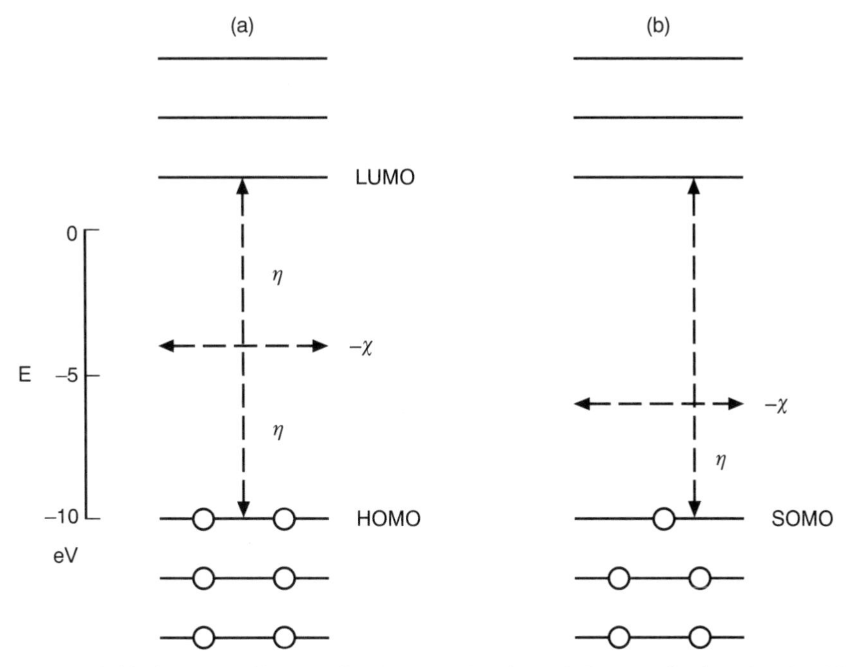

Figure 2.2 Orbital energy diagram for (a) a molecule and (b) a radical. HOMO = highest occupied molecular orbital; LUMO = lowest unoccupied MO; SOMO = singly occupied MO

This statement agrees with the earlier, imprecise definition. For example, optical polarizability in quantum theory results from a mixing of suitable excited-state wave functions with the ground-state wave function. The mixing coefficient is inversely proportional to the excitation energy from the ground to the excited state. A small HOMO–LUMO gap automatically means small excitation energies to the manifold of excited states. Therefore, soft molecules, with a small gap, will be more polarizable than hard molecules. High polarizability was the most characteristic property attributed to soft acids and bases.

A number of papers have appeared showing a correlation between polarizability, α, and softness.[17] Empirically it is found that $\alpha^{1/3}$ is a linear function of $(I - A)^{-1}$, the softness. This is equivalent to the classical result for spheres of radius R, that charging energies are proportional to 1/R, whereas polarizability is proportional to R^3. Calculations of α using DFT, and EN equalization, shows that α is equal to $(I - A)^{-1}$ times a factor dependent on the size of the system.

In simple MO theory, the energy gap between the HOMO and the LUMO defines the first excited state, or the energy of the first absorption band in the visible–UV. A hard molecule would have a large value of $h\nu_{max}$, where ν_{max} is the vertical frequency of the lowest-energy transition where the multiplicity does no change. For example, H_2O, H_2S, H_2Se and H_2Te have ν_{max} at 1655, 1950, 1970 and 2000 A, respectively, showing increasing softness.

It would be convenient if we could find $(I - A)$ from absorption spectra. But this is not possible, since $(I - A)$ is usually about twice as large as $h\nu_{max}$. For example, H_2S has $(I - A) = 12.6\,eV$, determined experimentally whereas $h\nu_{max} = 6.4\,eV$. The difference arises from the additional electron–electron repulsion that results from adding an electron to the LUMO, instead of merely exciting it from the HOMO. For filled-shell molecules,

$$(I - A) = h\nu_{max} + J_{HL} - 2K_{HL} \tag{2.18}$$

where J_{HL} and K_{HL} are coulomb and exchange integrals for one electron in the HOMO and in the LUMO.

Figure 2.3 shows a plot of $(I - A)$ for a number of molecules from Table 2.2 against $h\nu_{max}$ for the same molecules in the gas phase. The slope of the best straight line is 1.45. The scatter shows that molecules can have quite different values of J and K. In addition, there are errors due to orbital relaxations, not allowed for by Koopmans' theorem.

Table 2.4 gives the values for $h\nu_{max}$ for a number of selected molecules in the gas phase. Generally there is a good correlation between these energies and

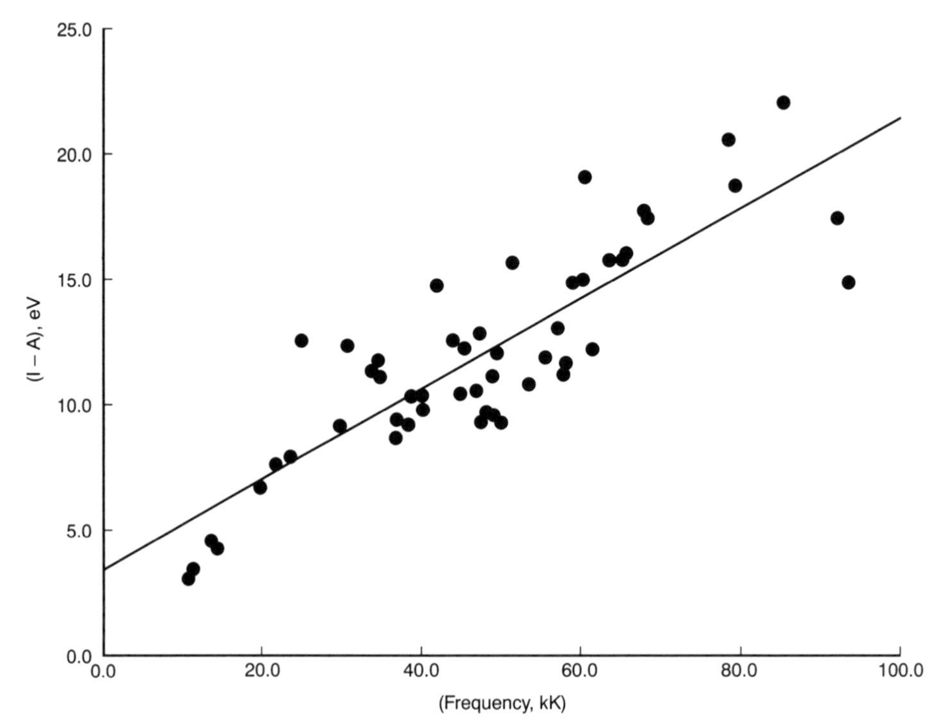

Figure 2.3 Plot of $(I - A)$ against the vertical frequency of the first absorption band for a number of gas-phase molecules. The multiplicity of the excited state is the same as that of the ground state. From [55], reprinted with permission.

Table 2.4 Vertical Frequencies of the First Absorption Band of Selected Molecules in the Vapor Phase[a]

Molecule	ν_{max} $[10^3\,cm^{-1}]$	Molecule	ν_{max} $[10^3\,cm^{-1}]$
CH_4	78.2	CH_3CN	60.0
C_2H_6	75.8	$CH_2=CHCN$	47.5
n-C_5H_{12}	64.0	$CH_2=CO$	25.9
i-C_5H_{12}	64.0	$CH_2=N_2$	21.2
CH_3F	75.4	HCOOH	45.5
CH_3Cl	59.0	$HCOOCH_3$	47.5
CH_3Br	50.0	CH_3COOH	47.5
CH_3I	38.5	$HCONH_2$	45.5
t-C_4H_9I	37.0	HCOF	45.5
CH_3OH	54.5	CH_3COCl	42.6
C_2H_5OH	55.0	CH_3NO_2	37.0
t-C_4H_9OH	55.0	C_2H_2	57.0
$(CH_3)_2O$	54.5	C_2H_4	61.4
THF	50.0	C_3H_6	58.0
HCHO	31.0	$(CH_3)_2C=C(CH_3)_2$	53.5
CH_3CHO	34.0	$CH_2=CHCH=CH_2$	48.3
CH_3COCH_3	35.0	C_6H_6	40.0
$CH_2=CHCHO$	26.5	CF_4	110.5
C_6H_5COOH	36.0	BF_3	63.6
C_6H_5OH	37.00	BCl_3	57.9
$C_6H_5NO_2$	37.00	$Cr(CO)_6$	31.5
C_5H_5N	40.3	RuO_4	26.0
H_2O	60.4	OsO_4	34.0
CO_2	68.0	BH_2	13.0
NH_3	51.5	NH_2	15.4
$(CH_3)_3N$	44.2	HCO	15.2
PH_3	55.6	NCO	25.0
H_2S	50.0	NO_3	17.2
$(CH_3)_2S$	43.9	CH_3	46.3
SF_6	93.0		

[a] Same multiplicity as ground state. data from Reference 18. 1000 cm^{-1} is one kiloKayser, kK.

$(I - A)$ for a series of related molecules, but some exceptions are seen. It would appear that CF_4 is the hardest molecule known, since $h\nu_{max}$ is 110.5 kK, compared with 78.2 for CH_4. Neither I or A is known accurately for CF_4, but I is about 16 eV, and A is about −8 eV, making $\eta \simeq 12$ eV, compared with 10.3 eV for CH_4.

The visible range is from 12.5 to 25 kK ($10^3\,cm^{-1}$). Some free radicals are included in Table 2.4, to remind us that many would be colored, if we could see them. While CH_3 is colorless, it is shifted towards the visible, compared with CH_4. These results are consistent with the great reactivity of free radicals.

THE FUKUI FUNCTION AND LOCAL HARDNESS

Since μ is a function of N and v, we can write the exact differential

$$d\mu = 2\eta dN + \langle f dv \rangle \tag{2.19}$$

where the Fukui function, f is defined as[20]

$$f = (\delta\mu/\delta v)_N = (\partial\rho/\partial N)_v \tag{2.20}$$

The last equality is a Maxwell relation based on Equation (2.8).[19]

In a chemical reaction between C and D, we can assume that changes in μ are a measure of the extent of reaction. In Equation (2.19), the first term on the right-hand side involves global quantities, and is not direction-sensitive. The Fukui function, however, is a local quantity and has different values at different parts of both molecules. Therefore the preferred orientation of C and D is that with the largest values of f at the reaction site.[20] This will lead to the largest value of $d\mu$.

We now adopt an MO viewpoint and use the "frozen core" assumption underlying Koopmans' theorem. The f can be identified with the electron density of the orbital involved. For an electrophilic attack of D on C, we have

$$f^- = \rho_{HOMO} \qquad \mu_C > \mu_D \tag{2.21}$$

and for a nucleophilic attack

$$f^+ = \rho_{LUMO} \qquad \mu_C < \mu_D \tag{2.22}$$

Finally, for a radical attack, or any case where electrons flow in both directions, we have

$$f_0 = \tfrac{1}{2}(\rho_{HOMO} + \rho_{LUMO}) \qquad \mu_C \simeq \mu_D \tag{2.23}$$

The orbitals are those of C, in all cases, but there would be complementary equations for D.

These are just the rules for classical frontier orbital theory, as proposed by Fukui[21] – hence the name Fukui function. DFT has again been reconciled with MO theory, this time for chemical reaction. The Fukui functions are reactivity indices. Further analysis of the interaction of C and D shows that the value of

$\langle f_C f_D \rangle$ helps determine the net change in chemical potential.[22] This implies that good orbital overlap is also required.

The necessity for three reactivity indices, as shown in Equations (2.21)–(2.23), reminds us that $(\partial \rho / \partial N)$ is not a continuous function but has discontinuities at integral values of N. The same is true for the energy, of course.[23] If our system consists of a single isolated molecule, then there is a constraint in that electrons may be lost, but not gained. In such a case μ is simply equal to the orbital energy of the HOMO.[24] The hardness is not defined.

Unlike the chemical potential, which must be the same everywhere in the system at equilibrium, the hardness is not required to be constant, and has local values, $\tilde{\eta}$. From Equations (2.7) and (2.12) we find that[25]

$$2\tilde{\eta} = \langle (\partial^2 F / \partial \rho \partial \rho') \rho / N \rangle \tag{2.24}$$

An average of these local values over the system then gives the global value, η. The calculation of $\tilde{\eta}$ is not easy, but Equation (2.24) is useful in telling us that the hardness depends only on the way in which the kinetic energy and the inter-electronic repulsion energy change with the number of electrons.

An easier function to deal with is the local softness, $\tilde{\sigma}$. This is defined as[26]

$$\tilde{\sigma} = (\partial \rho / \partial \mu)_v = (\partial \rho / \partial N)_v (\partial N / \partial \mu)_v = f \sigma \tag{2.25}$$

Thus the local softness is readily found from the global value by multiplying by the appropriate value of the Fukui function from equations (2.21)–(2.23). The local softness measures how easy it is to change the electron density at different parts of the molecule. As expected, the softness is different for accepting or losing electrons. Unfortunately, $\tilde{\eta}$ is not the simple inverse of $\tilde{\sigma}$.[27] However, it is reason able to assume that in comparing two sites in a molecule with quite different values of $\tilde{\sigma}$, that the $\tilde{\eta}$ values will also be quite different in the inverse sense.

As the positive charge on an atom increases, the electron density around it is compacted. It becomes more difficult to change and the site becomes harder. The increase in the classical Coulomb potential of the electrons is the main factor in this hardening.[28] The effect will be largest for points close to the nucleus, but will be appreciable at bonding distances. Local hardness will be an important property in determining bond strengths.

CHEMICAL REACTIVITY

The ultimate goal of any general theory of chemistry must be to give information about the relative stabilities of molecules, and their tendencies to undergo chemical change under specified conditions. DFT gives such information in detail when used to make *ab-initio* calculations, as described earlier. We will not dwell on this application, but instead look for more qualitative information from the concepts derived from DFT.

We can immediately draw important conclusions about molecular stability from Figure 2.2, and the identification of 2η with the HOMO–LUMO energy gap. Soft molecules will be less stable than similar hard molecules. They will dissociate or isomerize more readily. In the perturbation theory of such reactions, change occurs by mixing in excited-state wave functions with the ground-state wave function. If Q is the reaction coordinate,

$$\psi = \psi_0 + \sum_k \frac{\langle \psi_0 | (\partial E/\partial Q) | \psi_k \rangle}{(E_0 - E_k)} \psi_k \Delta Q \tag{2.26}$$

The index k refers to the excited states. Only states of the same multiplicity can mix.

There are important symmetry restrictions contained in Equation (2.26).[29] These are not important at this time, but what *is* important is that an easy change of ψ, and hence of ρ, can only occur if the transition energies $(E_0 - E_k)$ are small. There is a sum over all excited states in Equation (2.26), but only a few will be important in driving the reaction. These will usually lie close to the first excited state.

Figure 2.4(a) helps to explain this. It shows the ground state and some of the excited states on an energy scale. There is a dense manifold of excited states, some chemically important and others not, but $h\nu_{max}$ sets a lower limit to all of these. Also the difference $(h\nu_{max})$ will usually be much larger than the differences between the excited states. Figure 2.4(b) shows an exception in which there are

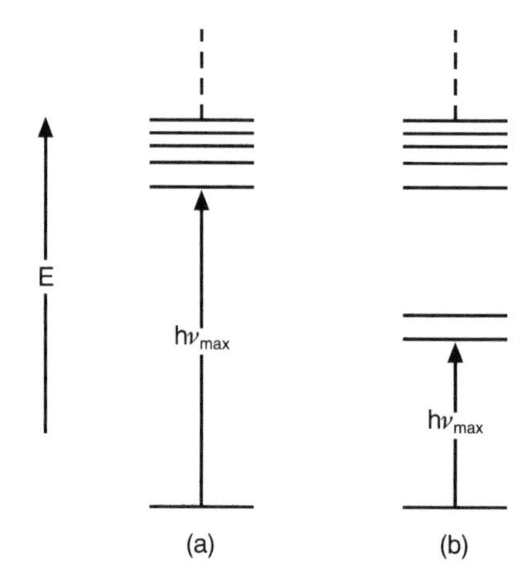

Figure 2.4 (a) Stable molecule with large energy gap between ground and excited states. (b) Stable molecule with small energy gap for excited states that are not chemically useful.

low-lying excited states which are not chemically important. An example would be the colored complexes of the transition metals.

As predicted from the above, there is a close connection between dissociation energies and $h\nu_{max}$.[18] Plots of the latter vs. bond dissociation energy D_0 for diatomic molecules are fairly linear, providing that covalent and ionic molecules are separated. In view of Figure 2.3, we can expect similar results for $(I - A)$ vs. D_0. Usually, if we replace a light atom in a molecule with a heavier member of the same family, we can expect two changes. The experimental value of $(I - A)$ will become smaller,[30] and the energy needed to break a bond to the replaced atom will become less.

There is another way to test chemical stability hat can be applied to aromatic organic compounds.[31] Aromaticity is usually taken to mean high stability and low reactivity. It can be quantified by calculating the resonance energy per electron (REPE). These values are usually calculated by simple Hückel theory, but they correlate well with experimental measures of reactivity.[32]

Figure 2.5 shows experimental values of $(I - A)/2$ for a series of benzenoid hydrocarbons plotted against their REPE. In the figure, benzene is the most stable compound and has the largest energy gap. Tetracene is the least stable and

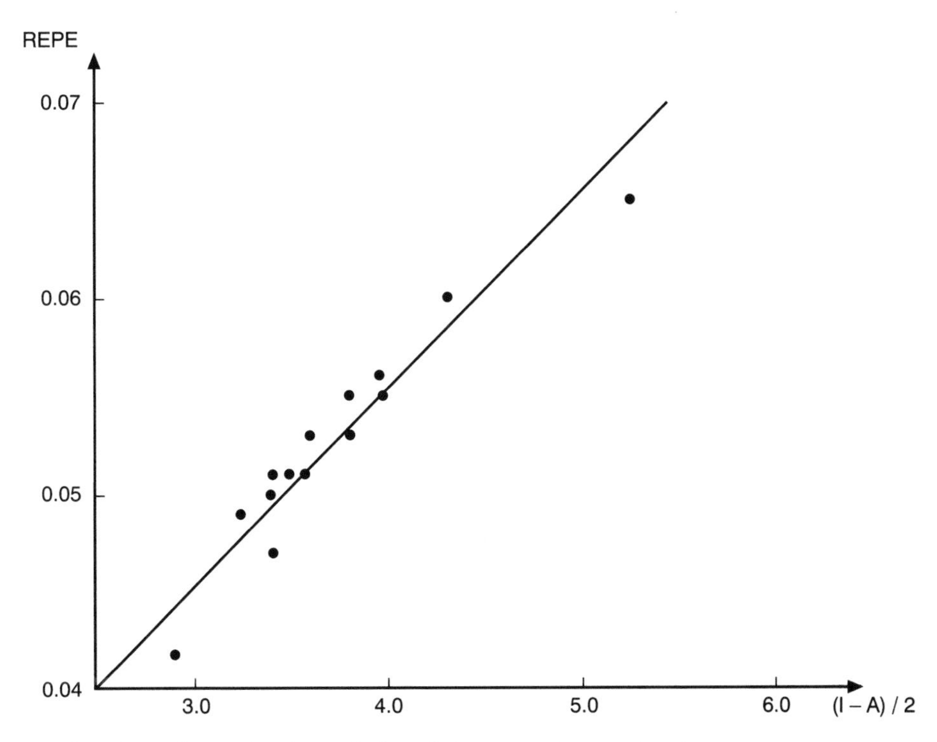

Figure 2.5 Experimental values of $(I - A)/2$ versus the calculated resonance energy per electron (REPE) for several benzenoid hydrocarbons

has the smallest gap. These results can be extended to many more examples, if we use η values also calculated from Hückel theory.[31]

We next consider hardness and its role in the case of bimolecular reactions

$$C + D \rightleftarrows CD^{\ddagger} \rightarrow \text{product} \tag{2.27}$$

where $CD^{\ddagger}$ is the transition state. We have already shown that there will be an initial transfer of electrons from D to C

$$\Delta N = \frac{(\chi_C^0 - \chi_D^0)}{2(\eta_C + \eta_D)} \tag{2.16}$$

This will result in an energy lowering:

$$\Delta E = -\frac{(\chi_C^0 - \chi_D^0)}{4(\eta_C + \eta_D)} \tag{2.28}$$

and a new value of the electronic chemical potential

$$\mu_{CD} = \frac{\mu_C^0 \sigma_C + \mu_D^0 \sigma_D}{(\sigma_C + \sigma_D)} \tag{2.29}$$

Equations (2.16) and (2.28) are very appealing, since they give information on reactivity using only data for the reactants, and a minimum number of parameters. In fact they were used before the development of DFT, mainly to estimate the polarity of bonds.[33] But they are obviously incomplete. They are based on C and D transferring electron density at distances that are so large that other interactions between them are negligible. However, as they come to bonding distance, there will be large changes in energy due to the added potentials of the nuclei.[34] Also, there will be changes due to the delocalization of electron density corresponding to the formation of covalent bonds.[35]

The energy lowering in Equation (2.28) is typically only a few kilocalories per mole, far short of bond energies. Bringing C and D closer together to form $CD^{\ddagger}$ creates interactions which require consideration of orbitals in MO theory, or Fukui functions in DFT. In the former case perturbation theory is generally used, as pioneered by Dewar[36] and Fukui and Fujimoto.[37] Figure 2.6 shows the MO–energy diagrams for two interacting molecules, which are of similar EN, so that electrons will flow in both directions. One molecule, however, is harder than the other.

Two kinds of interactions are shown:

1. Partial transfer of electrons from the HOMO of each molecule to the LUMO of the other. This occurs by mixing of the orbitals.
2. There is a mixing of the filled MOs of each molecule with its own empty MOs.

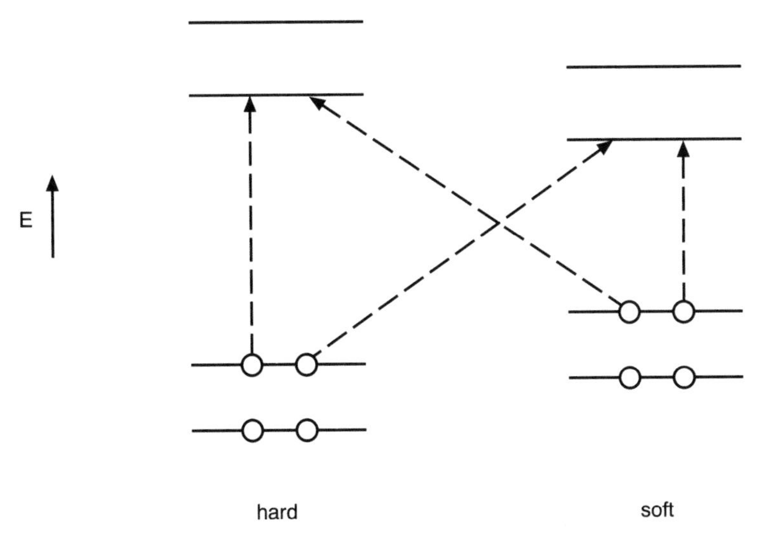

Figure 2.6 Partial transfer of electrons from HOMOs of each of two molecules to LUMOs of the other (delocalization), and mixing of excited states within the same molecule (polarization).

The first effect is called delocalization and is the mechanism whereby new bonds are formed between the reactants and old bonds are broken. Other orbitals besides the HOMOs and LUMOs may be involved to a lesser extent.

The second effect is called polarization. It has the effect of lowering the repulsive energy between the two molecules as they approach each other. Polarization is easiest when the energy gap is small for each molecule.

Considering only the four frontier orbitals, an approximate value for the energy lowering due to delocalization is given by

$$\Delta E = \frac{2\beta_1^2}{(A_C - I_D)} + \frac{2\beta_2^2}{(A_D - I_C)} \tag{2.30}$$

The βs are exchange integrals of the perturbation Hamiltonian over the interacting MOs. One β could correspond to σ bonding, and the other to π bonding, for example. The stabilization is greatest if A is large for both molecules and I is small. This means that both energy gaps should be small or both molecules should be soft.

Note that Equation (2.30) already requires two more parameters than Equation (2.28) does. Also, to estimate β_1 and β_2, we need to select definite occupied and empty orbitals on both C and D. Finally, as C and D approach more closely, the orbitals of the original reactants will become so changed as a result of mixing that they will be useless for further calculations.

It is of interest that a different MO perturbation theory, using localized orbital interactions, projects out the local softness and the Fukui functions directly.[38]

Both of these can be evaluated for the different atoms in each molecule, and they define the localizability of the interaction. Instead of the global EN, a local EN is calculated for each atom and determines the amount of local electron transfer.

There have been several papers probing the further interaction of two molecules using DFT.[22,39] The focus is on the electron density of the combined system, and how it changes as reaction proceeds. The guiding principle is the constancy of the chemical potential over the system. As expected, some parts gain electron density, and other parts lose it. The calculation of actual energy changes is not easy. The importance of the Fukui function, or the equivalent local softness, $\tilde{\sigma}$, is confirmed. Reaction usually takes place at the two atoms with the largest values of f. But this is only true if there is considerable electron transfer between C and D. For reactions between two hard molecules, reaction is favored at atoms where f has minimal values.

All of these theoretical studies, whether by MO or DFT methods, provide support for the HSAB Principle, "Hard likes hard, and soft likes soft". This is easily seen in Equation (2.16). Obviously if one reactant easily loses electrons, it is best if the other reactant easily gains electrons. Support for the HSAB Principle also comes from *ab-initio* calculations of Hartree–Fock accuracy on combinations of hard and soft metal ions with hard and soft neutral ligands.[40]

The energy changes can be broken down into contributions of electrostatics, exchange repulsion, ligand polarization, ligand-to-cation charge transfer and electron correlation effects. As expected, charge transfer is greatest in the reaction of soft metal ions with soft bases, and least in the reaction of hard metal ions with hard bases. The increased stability of a soft–soft combination is largely due to increased correlation energy. This reduces electron–electron repulsion. It is essentially the same as the polarization energy shown in Figure 2.6.

While all of these studies are in accord with the HSAB Principle, none of them can be taken as a general proof. Indeed, a rigorous proof may be very difficult to formulate. However, there is one noteworthy attempt.[41] Write Equation (2.28) as

$$\Delta E = \Delta\Omega_{\mathrm{C}} + \Delta\Omega_{\mathrm{D}} \tag{2.31}$$

$$\Delta\Omega_{\mathrm{C}} = -\frac{(\mu_{\mathrm{D}}^0 - \mu_{\mathrm{C}}^0)^2 \eta_{\mathrm{C}}}{4(\eta_{\mathrm{D}} + \eta_{\mathrm{C}})^2} \qquad \Delta\Omega_{\mathrm{D}} = -\frac{(\mu_{\mathrm{D}}^0 - \mu_{\mathrm{C}}^0)^2 \eta_{\mathrm{D}}}{4(\eta_{\mathrm{D}} + \eta_{\mathrm{C}})^2} \tag{2.32}$$

Assume that for a given $(\mu_{\mathrm{D}}^0 - \mu_{\mathrm{C}}^0)$ and η_{D}, $\Delta\Omega_{\mathrm{C}}$ is minimized with respect to η_{C}. It follows that $\eta_{\mathrm{C}} = \eta_{\mathrm{D}}$. Minimizing $\Delta\Omega_{\mathrm{D}}$ will give the same result, so that $\Delta\Omega_{\mathrm{C}} = \Delta\Omega_{\mathrm{D}}$, and $(\Delta\Omega_{\mathrm{C}} + \Delta\Omega_{\mathrm{D}})$ is also minimized.

The reason for this procedure is that the product molecule, CD, can be considered an open system. That is, there is an exchange of both energy and matter (electrons) between the parts, *C* and *D*. In ordinary thermodynamics, the quantity that determines equilibrium in an open system is the grand potential, $\Omega = E - N\mu$. At equilibrium the grand potential is as negative as possible.[42] But the $\Delta\Omega$ values of Equation (2.32) are just the changes in their grand potentials. This assumes that a single molecule, where N is the number of electrons, can be considered in the same way as a collection of molecules, where N is their number.

But this is the same assumption that has already been made in the case of Equations (2.9) and (2.12), which define μ and η. Accordingly, we have proved that if a given acceptor molecule reacts with a set of donor molecules, the most stable product will be formed with the donor whose hardness is the same as that of the acceptor. The HSAB Principle! However there is an important restriction. The quantity $(\mu_C - \mu_D)$ must be constant, so that only donors of the same electronic chemical potential can be compared. A similar conclusion can be drawn for a series of acceptors.

This restriction is reminiscent of the restriction applied to Equation (1.7) of Chapter 1, that only acids (or bases) of comparable strength can be considered. It is not quite the same since, as explained earlier, orders of electronegativity are not orders of strength. The main objection to the above proof of the HSAB Principle is that it is based on Equation (2.28), which only gives the initial interaction of C and D. The fact that it apparently leads to the right result suggests that equations (2.28) and (2.16), while incomplete, may often be harbingers of better values of ΔE and ΔN.

ELECTRONEGATIVITY SCALES

The concept of electronegativity (EN) is almost as old as chemistry itself. Berzelius classified atoms as electronegative or electropositive. By the turn of the 20th century it was understood that these terms referred to the electron-attracting and -holding power of the atoms. During the 1920s the founders of physical-organic chemistry extended the terms to include groups of atoms as well as atoms. There was an approximate ordering of the EN of various atoms and radicals.

In 1932 Pauling made a landmark contribution.[43] He created an empirical scale of EN based on heats of formation or, essentially, bond energies. A number of other scales eventually appeared, such as the widely used Allred–Rochow scale.[44] These scales had two characteristics in common: one was that they were calculated from properties of the free atoms of the elements; the other was that they were tested by seeing if they agreed with the original Pauling scale. Failure to do so would be a serious deficiency. In 1939, in the first edition of *The Nature of the Chemical Bond*, Pauling gave his meaning of the word electronegativity: "the power of an atom in a molecule to attract electrons to itself". Many would accept this as a definition of the term.

It is clear that absolute EN differs substantially from Pauling EN. It applies to molecules, ions and radicals, as well as to atoms. For the latter, it is a property of a free atom in the ground state and not an atom in an excited valence state suitable for its appearing in a molecule. As might be expected, applications of the two scales are quite different. The Pauling scale is useful for estimating bond polarities and, to some degree, the strengths of bonds between different atoms. The absolute scale is a measure of the chemical reactivity of an atom, radical, ion or molecule.

The absolute EN does not conform to the Pauling definition of EN as a property of an atom in a molecule, but the essential idea of EN is that of attracting and holding electrons. There is no compelling reason to restrict this to combined atoms. The extension of the concept of EN to molecules seems to be a natural and useful step. Donor–acceptor interactions are at the very heart of chemical bonding. The absolute EN is a measure of the intrinsic donor–acceptor character of a species.

There is no inconsistency in the EN of a free atom being different from that of an atom in a valence state. Scales such as Mulliken's and the recently developed spectroscopic scale[45] show that the absolute and Pauling-like scales can be commensurable. Since the applications are so different, it is not a meaningful question to ask which scale is more correct. Each scale is more correct in its own area of use.

While Equation (2.16), in principle, can be used to calculate bond polarities, it is not as reliable as methods using the Pauling scale. However, the absolute scale of EN can be used in a unique way to probe bond polarity. For example, consider a molecule X—Y, consisting of two atoms or radicals held together by a bond. The polarity of the bond could determine whether the molecule behaves as X^+, Y^- or X^-, Y^+. The same question can be asked by looking at the reactions

$$X^-(g) + Y^+(g) = X\text{—}Y(g) = X^+(g) + Y^-(g) \tag{2.33}$$

The difference in energy between the products on the right and those on the left is easily found:

$$\Delta E = (I_X - A_Y) - (I_Y - A_X) = 2(\chi_X - \chi_Y) \tag{2.34}$$

If X has a greater absolute EN than Y, ΔE is positive. This means that X—Y acts as X^-, Y^+. The answer is a thermodynamic one and involves no assumptions.

Since the Pauling scale has no meaning for molecules, or even ions, we can only compare the absolute and Pauling scales for atoms and radicals. Table 2.5 gives the EN values for the more common elements as χ_A and χ_P, both in electron volts. The Pauling values are actually the spectroscopic ones of Allen, which are in remarkable agreement with the Pauling scale. They are the average ionization potentials of the valence shell of the atoms and are given in electron volts. The values for η are also given in the table.

The scales are roughly parallel, but there are definite deviations. For example, B and Al are less EN than Be and Mg on the absolute scale. This disagrees with our expectation that EN will increase smoothly as we go from left to right in any row of the Periodic Table. But this is by no means proof that the Pauling scale is

Table 2.5 χ_A and χ_P [eV]

Atom	$\chi_A{}^{(a)}$	$\chi_P{}^{(b)}$	$\eta^{(a)}$
H	7.18	13.61	6.43
Li	3.01	5.39	2.39
Be	4.9	9.32	4.5
B	4.29	12.13	4.0
C	6.27	15.05	5.00
N	7.30	18.13	7.23
O	7.54	21.36	6.08
F	10.41	24.80	7.01
Na	2.85	5.14	2.30
Mg	3.75	7.65	3.90
Al	3.23	9.54	2.77
Si	4.77	11.33	3.38
P	5.62	13.33	4.88
S	6.22	15.31	4.14
Cl	8.30	16.97	4.68
K	2.42	4.34	1.92
Ca	2.2	6.11	4.0
Sc	3.34	6.8	3.20
Ti	3.45	7.4	3.37
V	3.6	8.1	3.1
Cr	3.72	8.6	3.06
Mn	3.72	9.2	3.72
Fe	4.06	9.9	3.81
Co	4.3	10.4	3.6
Ni	4.40	11.0	3.25
Cu	4.48	10.8	3.25
Zn	4.45	9.39	4.94
Ga	3.2	10.39	2.9
Ge	4.6	11.80	3.4
As	5.3	13.08	4.5
Se	5.89	14.34	3.87
Br	7.59	15.88	4.22
Rb	2.34	4.18	1.85
Sr	2.0	5.70	3.7
Y	3.19	5.9	3.19
Zr	3.64	6.6	3.21
Nb	4.0	7.4	3.0
Mo	3.9	8.2	3.1
Ru	4.5	9.8	3.0
Rh	4.3	10.6	3.16
Pd	4.45	11.3	3.89

Table 2.5 *(continued)*.

Atom	χ_A[a]	χ_P[b]	η[a]
Ag	4.44	11.7	3.14
Cd	4.33	8.99	4.66
In	3.1	9.79	2.8
Sn	4.30	10.79	3.05
Sb	4.85	11.74	3.80
Te	5.49	12.76	3.52
I	6.76	13.95	3.69
Cs	2.18	3.89	1.71
Ba	2.4	5.21	2.9
La	3.1	–	2.6
Hf	3.8	–	3.0
Ta	4.11	–	3.79
W	4.40	–	3.58
Re	4.02	–	3.87
Os	4.9	–	3.8
Ir	5.4	–	3.8
Pt	5.4	–	3.2
Au	5.77	–	3.46
Hg	4.91	10.44	5.54
Tl	3.2	–	2.9
Pb	3.90	–	3.53
Bi	4.69	–	3.74

[a] R.G. Pearson, *Inorg. Chem.*, **27**, 730 (1988).
[b] Reference 45.

more correct. This scale tells us that bonds between non-metallic elements and Be or Mg are more ionic than bonds of B or Al. The absolute scale tells us that it is easier to remove a 2p or 3p electron than a 2s or 3s electron. Both statements are equally true.

Unfortunately the fact that the same label, electronegativity, is used for both scales creates ample opportunity for confusion and misunderstanding. Since the Pauling scale has the advantage of seniority and long-established usage, a solution may be to find another term for the absolute scale. One alternative is to use the name "electronic chemical potential", μ.

This presents some difficulties. The μ scale is a set of negative numbers, and it is always more difficult to decide which of two negative numbers is the larger! More serious is the conflict with the existing usage of μ as the ordinary thermodynamic chemical potential μ_T:

$$\mu_T = \left(\frac{\partial E}{\partial N}\right)_{S,V} = \left(\frac{\partial G}{\partial N}\right)_{P,T} \tag{2.35}$$

Table 2.6 Relationship between Energy and Electronic Chemical Potential[a]

	$-E$	$-N\mu$		$-E$	$-N\mu$
O	74.809	0.055	CH_3OH	114.936	7.796
Kr	2757.81	18.867	LiF	106.989	5.822

There is a relationship between μ and μ_T, but it is not a simple one. Recall that μ depends on constraints on the changes in the number of electrons. If the system is a single molecule, then an electron can be lost but not gained. In such a case $\mu = \varepsilon_{HOMO}$, as already mentioned.[24] The relationship between the energy of the molecule, E, and μ is then given by[8]

$$E = N\mu + F - \langle|\partial F/\partial\rho|\rho\rangle + V_{nn} \tag{2.36}$$

where F is the sum of the kinetic energy and the electron–electron repulsion energy, as before. It is an explicit, but unknown, functional of ρ. V_{nn} is the nuclear–nuclear repulsion.

For all molecules except the smallest, E is much larger than $N\mu$. Examples are given in Table 2.6. Most of the energy is not relevant to chemistry since it comes from the inner-shell electrons. For this reason, we subtract from E the energies of the constituent atoms. This gives us the so-called electronic energy of statistical thermodynamics, E_T, which is by far the largest part of the thermodynamic chemical potential.[46]

Table 2.7 shows some energy changes on forming stable molecules from the atoms. Now we see that changes in $N\mu$ are comparable with changes in total

Table 2.7 Energy [au] of some Molecules in Terms of Components[a]

	$-E_T$	$-\Delta N\mu$	ΔV_{ee}	ΔV_{nn}
H_2	0.134	0.171	0.677	0.714
H_2O	0.250	−0.987	9.405	9.239
LiF	0.147	−1.336	11.924	9.686
CO	0.288	0.110	21.720	22.514
CO_2	0.419	−0.705	63.512	62.622
CH_3OH	0.438	−1.859	40.256	40.189
C_2H_6	0.721	0.456	41.873	41.932
NaCl	0.116	−0.886	44.493	41.695
BeO	0.072	−1.560	13.723	13.127

[a]All quantities are the changes upon forming the molecule from its atoms, and all are based on near Hartree–Fock calculations. Energies in atomic units.

energy, E_T, since they both involve the valence-shell electrons. Also included for comparison are the changes in total repulsion energy, which are much larger. Not shown are the large changes in V_{ne}, the nuclear-electronic attraction which lead to the formation of the molecule.

While E_T is always negative in forming a stable molecule, $\Delta N\mu$ can be either positive or negative. This means that μ cannot be used to predict bond energies or equilibrium constants for chemical changes, even though it is a component of E_T and μ_T, which can be used make such predictions. Although μ is constrained to be constant at equilibrium, it is not required by DFT to have a minimum value.

In summary, unless a better name can be coined, it appears that the term "absolute EN" will be around for some time. This means that chemists must be aware of the difference between χ_P and χ_A, and make it clear which scale they have in mind.

Considering the relationship between the two scales further leads to an interesting result:[47] χ_A changes to χ_P as a result of changing the electron density about the atom due to chemical bond formation. But changes in ρ depend on the hardness of the atom in question. Perhaps some mix of χ_A and η for the free atom will lead to Pauling EN.

Figure 2.7 shows the result of searching for such a mix.[48] The equation of the straight line is

$$\chi_P = 0.44\eta + 0.044\chi_A + 0.04 \tag{2.37}$$

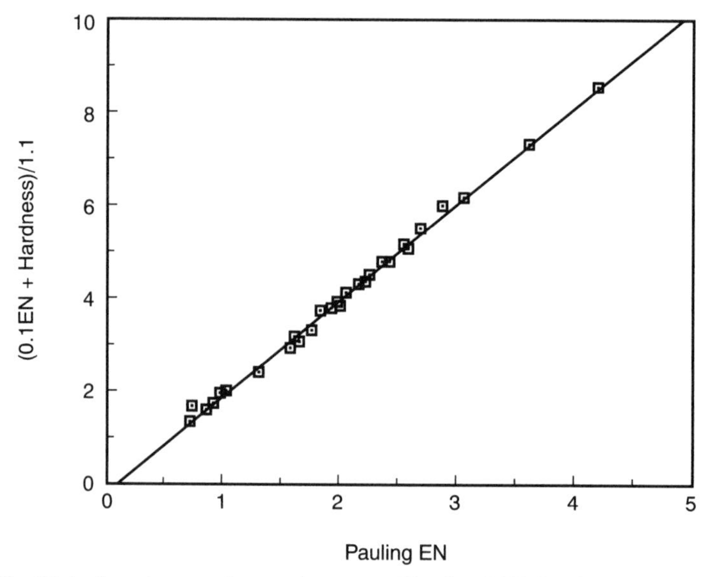

Figure 2.7 Plot of a mixture of χ_A and η vs. χ_P (Pauling EN) for the representative elements. After Reference 48.

Thus the Pauling scale is more closely related to hardness than to the absolute EN. This is perhaps not too surprising, if we remember the original concept of a hard acid or base as one where the acceptor or donor atom held on to its valence-shell electrons tightly.

However it should be pointed out that Figure 2.7 shows only the representative element atoms. The results would not be so good if the transition metals were included. Also, calculated values of η are shown, and not the experimental ones. Nevertheless, the close relationship of χ_P to η is undeniable. As might be expected, there is a linear relationship between $\alpha^{-1/3}$ and χ_P,[48] just as there is between $\alpha^{1/3}$ and η^{-1}.

CALCULATED μ AND η VALUES

When experimental values of I and A are lacking, calculated results are often satisfactory. A number of methods can be used, including both Hartree–Fock and density functional theory.[51] The best method would be an accurate calculation of the energies of M, M^+ and M^-, where M is a molecule. This gives reliable values when M^- is a stable species, i.e., for positive electron affinities.

Negative electron affinities are troublesome, because the variational method will simply give $A = 0$. This is the adiabatic value with the electron not attached to M. Fortunately, the self-consistent calculations, either by HF or KS methods, give the orbital energies of the empty, or virtual, orbitals as well as the occupied ones. The energies of the latter are for an electron in the field of $(N - 1)$ electrons, which is correct. The virtual orbital energies are for a fictitious electron in the field of N electrons.

This makes the calculations poor for matching electronic spectra, but it is just right for the electron affinity. There is still a problem, however, because the LUMO energy is very dependent on the size of the basis set used, and its quality. This means that results for both μ and η can be seriously in error. Fortunately, it appears that relative values for a series of related molecules, or a series of possible structures for a given molecule, are often quite reliable. The same basis set, or its equivalent, must be used.

It is also possible to calculate μ and η in DFT by using the fundamental definitions of Equations (2.9) and (2.12). The method used is the transition-state procedure introduced by Slater.[52] In essence, it requires the assumption that the energy is a continuous function of N, so that derivatives such as Equations (2.9) and (2.12) have exact meanings. Although it is not true in reality, this is a convenient mathematical device which gives useful results. Actually what is done is to calculate both μ and I by the transition-state method, and then to find A and hence η, by Equations (2.11) and (2.13).

Table 2.8 Orbital Energies [au] for Argon

Orbital	HF energy[(a)]	KS energy[(b)]	Experimental[(c)]
1s	118.61	114.41	117.82
2s	12.32	11.11	11.99
2p	9.57	8.73	9.17[d]
3s	1.22	1.07	1.08
3p	0.59	0.56	0.58

[(a)] E. Clementi and C. Roetti, At. Data Nucl. data Tables, **14**, 177 (1974).
[(b)] Q. Zhao, R.C. Morrison and R.G. Parr, Phys. Rev. A, *50*, 2138 (1994).
[(c)] K. Siegbahn *et al.* ESCA Applied to Free Molecules, North Holland/ American Elsevier, New York, 1969.
[(d)] Average of $J = 1/2$ and $3/2$.

There are, in fact, two cases where a fractional value of N makes good sense. One is a statistical mixture of molecules where the average value of N need not be integral.[23] The other is exemplified by Equation (2.16), where molecules C and D, after their initial reaction, will have non-integral N_C and N_D values. In molecules, in general, chemists have long considered that a population analysis will give N values for each atom that are not integers. This is the basis for the concept of electronegativity, after all.[53]

Since all of the above calculations are strongly dependent on orbital energies, it is worthwhile to close with a short comparison of orbital energies, as calculated by HF and by DFT, and as measured experimentally by ESCA and photoelectron spectroscopy. These are shown in Table 2.8. Both the HF and KS orbital energies are quite close to the experimental ionization potentials. In principle, the KS energy for the outermost orbital should equal the first ionization potential, but this has not yet happened. It will be recalled that the KS results depend on how well the exchange-correlation potential is represented.[54]

REFERENCES

1. General references on DFT: R.G. Parr and W. Yang, *Density Functional Theory for Atoms and Molecules*, Oxford University Press, New York, 1989; N.H. March, *Electron Density Theory of Atoms and Molecules*, Academic Press, New York, 1992; E.S. Kryachko and E.V. Ludema, *Density Functional Theory of Many Electron Systems*, Kluwer Press, Dordrecht, 1990.
2. P. Hohenberg and W. Kohn, *Phys. Rev.*, **136**, B864 (1964).
3. For excellent brief reviews of DFT, see T. Ziegler, *Chem. Rev.*, **91**, 651 (1991); W. Kohn, A.D. Becke and R.G. Parr, *J. Phys. Chem.*, **100**, 12974 (1996).
4. J.C. Slater, *Phys. Rev.*, **81**, 385 (1951); J.C. Slater, *Adv. Quantum Chem.*, **6**, 1 (1972); J.C. Slater and K.H. Johnson, *Phys. Rev.*, **135**, 544 (1972).

5. J.G. Snijders and E.J. Baerends, *J. Mol. Phys.*, **36**, 1789 (1978).
6. L. Fan and T. Ziegler, *J. Am. Chem. Soc.*, **114**, 10890 (1992) N. Allinger and K. Sakakibara, *J. Phys. Chem.*, **99**, 9603 (1995).
7. W. Kohn and L.J. Sham, *Phys. Rev.*, **140**, A1133 (1965).
8. R.G. Parr, R.A. Donnelly, M. Levy and W.E. Palke, *J. Chem. Phys.*, **68**, 3801 (1978).
9. M. Levy and J.P. Perdew, *Phys. Rev. A*, **32**, 2010 (1985).
10. R.S. Mulliken, *J. Chem. Phys.*, **2**, 782 (1934).
11. R.P. Iczkowski and J.L. Margrave, *J. Am. Chem. Soc.*, **83**, 3547 (1961).
12. R.T. Sanderson, *Science*, **121**, 207 (1955).
13. R.G. Parr and R.G. Pearson, *J. Am. Chem. Soc.*, **105**, 7512 (1983).
14. G.J. Schulz, *Phys. Rev.*, **5**, A1672 (1972); K.D. Jordan and P.D. Burrow, *Chem. Rev.*, **87**, 557 (1987).
15. R.G. Pearson, *Proc. Nat. Acad. Sci. USA*, **83**, 8440 (1986).
16. T. Koopmans, *Physica*, **1**, 104 (1934).
17. L. Komorowski, *Chem. Phys.*, **114**, 55 (1987); T.K. Ghanty and S.K. Ghosh, *J. Phys. Chem.*, **97**, 4951 (1993); U. Dinur, ibid., 7894; S. Hati and D. Datta, *J. Phys. Chem.*, **98**, 1436 (1994).
18. R.G. Pearson, *J. Am. Chem. Soc.*, **110**, 2092 (1988).
19. R. Nalewajski and R.G. Parr, *J. Chem. Phys.*, **77**, 399 (1982).
20. R.G. Parr and W. Yang, *J. Am. Chem. Soc.*, **106**, 4049 (1984).
21. K. Fukui, *Theory of Orientation and Stereoselection*, Springer-Verlag, Berlin, 1972; I. Fleming, *Frontier Orbitals and Organic Chemical Reactivity*, John Wiley, New York, 1976.
22. M. Berkowitz, *J. Am. Chem. Soc.*, **109**, 4823 (1987).
23. J. P. Perdew, R.G. Parr, M. Levy and J.L. Balduz, Jr., *Phys. Rev. Lett.*, **49**, 1691 (1982).
24. J.A. Alonso and N.H. March, *J. Chem. Phys.*, **78**, 1382 (1983).
25. S.K. Ghosh and M. Berkowitz, *J. Chem. Phys.*, **83**, 2976 (1985).
26. W. Yang and R.G. Parr, *Proc. Nat. Acad. Sci. USA*, **82**, 6723 (1985).
27. M. Berkowitz and R.G. Parr, *J. Chem. Phys.*, **88**, 2554 (1988).
28. M. Berkowitz, S.K. Ghosh and R.G. Parr, *J. Am. Chem. Soc.*, **107**, 6811 (1985).
29. This is called the second-order Jahn–Teller (SOJT) method; R.G. Pearson, *Symmetry Rules for Chemical Reactions*, John Wiley, New York, 1976.
30. W. Yang, C. Lee and S.K. Ghosh, *J. Phys. Chem.*, **89**, 5412 (1985).
31. Z. Zhou and R.G. Parr, *J. Am. Chem. Soc.*, **111**, 7371 (1989).
32. Z. Zhou, *Int. Rev. Phys. Org. Chem.*, **11**, 243 (1992).
33. J. Hinze, M.A. Whitehead and H.H. Jaffe, *J. Am. Chem. Soc.*, **85**, 148 (1963); R.S. Evans and J.E. Huheey, *J. Inorg. Nucl. Chem.*, **32**, 373 (1970).
34. R.F. Nalewajski, *J. Am. Chem. Soc.*, **106**, 944 (1984).
35. R.G. Pearson, *J. Am. Chem. Soc.*, **107**, 6801 (1985).
36. M.J.S. Dewar, *J. Am. Chem. Soc.*, **74**, 3341, 3357 (1952).
37. K. Fukui and H. Fujimoto, *Bull. Chem. Soc. Japan*, **41**, 1989 (1968); idem, ibid., **42**, 3399 (1969).
38. H. Fujimoto and S. Satoh, *J. Phys. Chem.*, **98**, 1436 (1994).
39. A. Tachibana and R.G. Parr, *Int. J. Quantum Chem.*, **41**, 527 (1992); J.L. Gázquez and F. Méndez, *J. Phys. Chem.*, **98**, 4591 (1994); Y. Li and J.N.S. Evans, *J. Am. Chem. Soc.*, **117**, 7756 (1995).
40. P.K. Chattaraj and P.v.R. Schleyer, *J. Am. Chem. Soc.*, **116**, 1067 (1994); D.R. Garmer and N. Gresh, ibid., 3556.
41. P.K. Chattaraj, H. Lee and R.G. Parr, *J. Am. Chem. Soc.*, **113**, 1855 (1991).

42. See properties of the grand canonical ensemble, for example, in T.L. Hill, *Statistical Mechanics*, McGraw-Hill, New York, 1956, p. 72.
43. L. Pauling, *J. Am. Chem. Soc.*, **54**, 3570 (1932).
44. For a summary of scales see J. Mullay, in *Structure and Bonding*, 66 "Electronegativity", 1 (1987).
45. L.C. Alen, *J. Am. Chem. Soc.*, **111**, 9003 (1989); L.C. Allen and E.T. Knight, *J. Mol. Struct. (Theochem.)*, **261**, 313 (1992).
46. The zero point energy is usually added to E_T, making it less negative.
47. L. Komorowski, *Chem. Phys. Lett.*, **103**, 201 (1983); *Z. Naturforsch*, **42A**, 767 (1987).
48. J.K. Nagle, *J. Am. Chem. Soc.*, **112**, 4741 (1990); J.K. Nagle, private communication.
49. R.G. Pearson, *J. Orgn. Chem.*, **54**, 1423 (1989).
50. S.G. Lias, L.F. Liebman and R.D. Levin, *J. Phys. Chem. Ref. Data*, **17**, Suppl. No. 1 (1988).
51. See the papers in *Structure and Bonding*: (a) 66 "Electronegativity", K.D. Sen and C.K. Jorgensen, Eds. (1987); (b) 80 "Chemical Hardness", K.D. Sen, Ed. (1993).
52. J.C. Slater, *Quantum Theory of Molecules and Solids*, McGraw-Hill, New York, 1974, Vol. 4.
53. For an informative review see D. Bergmann and J. Hinze, *Angew. Chem., Int. Ed. Engl.*, **35**, 150 (1996).
54. For a discussion see A.D. Becke, *J. Chem. Phys.*, **98**, 5648 (1993).
55. J.B. Maksić, (Editor) *Theoretical Models of Chemical Bonding, Part 2*. Springer-Verlag, Heidelberg, 1990, p. 55.

3 Application of DFT

INTRODUCTION

In this chapter we will give examples of the uses of DFT in understanding chemistry. The emphasis will be on the concepts, namely EN and hardness, as given by DFT. In addition, there is the use of *ab-initio* DFT calculations, but these will only be mentioned when they help to understand the concepts.

Figure 3.1 shows the usual MO energy diagrams for several molecules. This is a convenient way of assessing a chemical species in cases where I and A are known. The direction of spontaneous electron flow will be from $(CH_3)_2O$ to Mg, and from Mg, or $(CH_3)_2O$, to Cl_2. The ether is a hard molecule, which limits the amount of electron transfer, whereas Mg and Cl_2 are soft. The amount of initial transfer is given by the equation

$$\Delta N = \frac{(\chi_C - \chi_D)}{2(\eta_C + \eta_D)} \tag{3.1}$$

This equation can be very useful, as we shall show. However, it is easy to overestimate its importance. Chemical reaction between two species depends on many other factors besides the ease of electron transfer; the donating and accepting orbitals are certainly among these other factors to be considered. But equally important is the nature of the reactants, which determines whether there is a reasonable reaction path to reach stable products. This will be illustrated by considering reaction between Cl_2 and both Mg and $(CH_3)_2O$ in detail.

Calculation with Equation (3.1) gives the results

$$Mg \rightarrow Cl_2 \qquad \Delta N = 0.05\,e^- \tag{3.2}$$

$$(CH_3)_2O \rightarrow Cl_2 \qquad \Delta N = 0.10\,e^- \tag{3.3}$$

which does not explain why Equation (3.2) represents an energetic reaction leading to $MgCl_2$, whereas reaction (3.3) is only mildly energetic, leading to the charge transfer complex $(CH_3)_2O{:}Cl_2$. Consideration of the orbitals involved is of little help. In both cases the electrons come from an orbital which is nonbonding, a b_1 orbital in the case of the ether, and an s orbital in the case of the metal.[1] The accepting orbital on Cl_2 is an anti-bonding σ_u.

Putting electron density into the σ_u orbital helps break the Cl–Cl bond.[2] But the bonding in the ether remains intact and the reaction stops after the initial charge

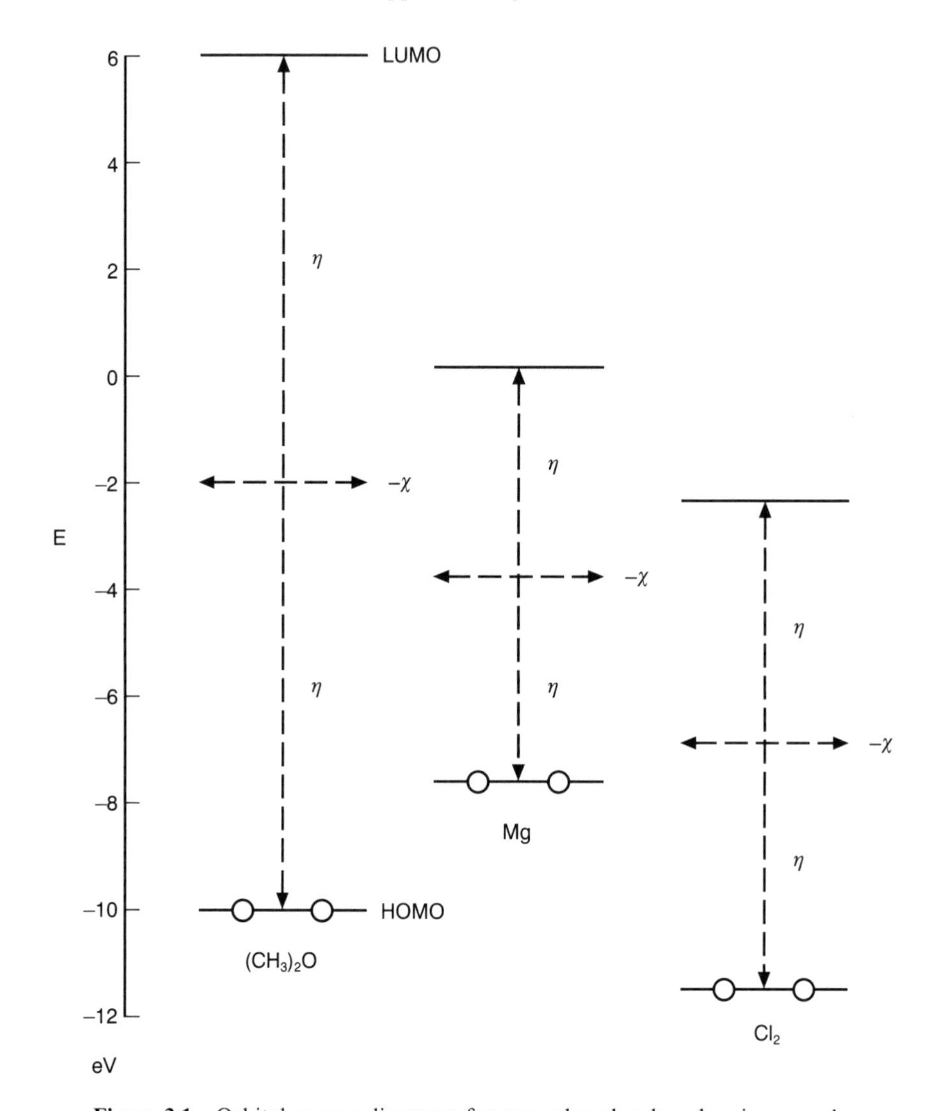

Figure 3.1 Orbital energy diagrams for several molecules, showing χ and η.

transfer. In the case of the atom, however, we can go on to a molecule of $MgCl_2$, usually considered to be about 75% ionic,[3] by an easy sequence of events:

$$Mg + Cl_2 \rightarrow MgCl^+,Cl^- \rightarrow \overset{-0.75\,+\,1.5\,-\,0.75}{Cl{-}Mg{-}Cl} \tag{3.4}$$

The last step is a simple rotation. The ether does not follow the same path because of the higher energy of both $(CH_3)_2OCl^+,Cl^-$ and $(CH_3)_2OCl_2$, containing tetracovalent oxygen.

Figure 3.2 shows the HOMO, the LUMO and χ for several more molecules. NH_3 will donate electrons to Pd, and Pd will donate electrons to CO. Experimentally, Pd:NH_3 is an unstable compound, Pd:CO is more stable, and NH_3:CO shows no signs of existing, though NH_3 should donate electrons to CO more than Pd does. In this case, consideration of the orbitals does help to understand the results.

Figure 3.3(a) shows the accepting orbital (the LUMO) of CO. The donating orbital (not shown) is a filled d orbital on Pd, or any transition metal, M. Clearly there is good overlap between the two orbitals. Also not shown are the HOMO of CO, a weakly anti-bonding σ orbital concentrated on carbon, and the LUMO of Pd, the 5s orbital. These also overlap well and lead to the synergistic effect of

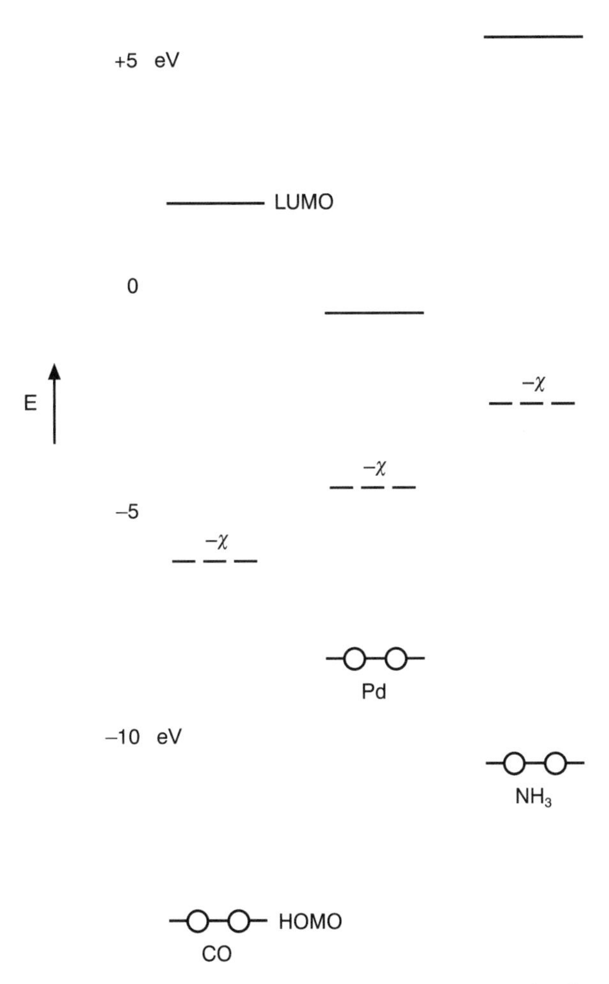

Figure 3.2 Orbital energy diagram showing HOMO, LUMO and χ for several species.

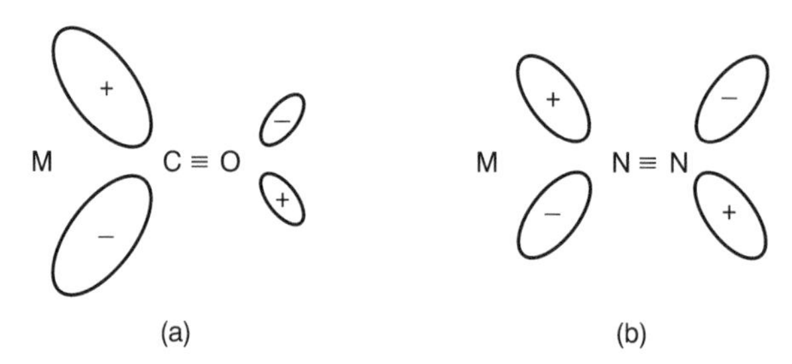

Figure 3.3 The π^* orbitals of (a) CO (b) N_2, showing greater overlap of CO with a d orbital on the transition metal atom, M.

$(\pi + \sigma)$-bonding. NH_3, being a pure σ-donor, cannot interact in this way with either Pd or CO. The empty π orbitals of NH_3 are very high in energy, even higher than the LUMO shown.

The conclusion that π-bonding in metal carbonyls is more extensive than σ-bonding was at first quite surprising.[4] Looking at the formula $M(CO)_n$, most chemists would consider the metal atom center as the acceptor and the CO ligands as the electron donors. In fact CO was earlier listed as a soft base. It now must be considered an acid, and a hard one at that!

Earlier *ab-initio* calculations on metal carbonyls were contradictory on the question of whether π-bonding or σ-bonding was dominant. However the latest, and best, calculations using DFT with relativistic corrections show clearly that π-bonding is the more important.[5] There is a wealth of information available on bond strengths and stabilities of transition metal carbonyls. Since we are dealing with orbitals on CO and the metal that are rather constant, we can check to see if Equation (3.1) is consistent with the bond energy data for various metals.

There is a problem in that the ground state of the metal atoms are all high-spin, except for Pd, which is $(4d)^{10}$. The metal carbonyls are all low-spin with all electrons paired up in d orbitals, as far as possible. Fortunately, in most cases it is possible to correct I and A for this change in configuration, using spectroscopic data.[4] Although DFT is not really valid for these excited states, we can assume that a low-spin metal atom, M, is a good model for a low-spin $M(CO)_{n-1}$, which then reacts with CO to form $M(CO)_n$.

Table 3.1 gives a list of χ and η values calculated for the valence state of the transition metals where the data are available. The valence state is $(nd)^k$ in all cases, except Cu, Ag and Au, where the ground state and valence state are both $(nd)^{10}(n+1)s$, but ionization is from the d shell and not the s. The values of ΔN calculated from Equation (3.1) are also given, and the values of $\Delta H^\ddagger$ for the reaction

$$M(CO)_n(g) = M(CO)_{n-1}(g) + CO(g)\ \Delta H^\ddagger \tag{3.5}$$

from experiment, where known.

Table 3.1 Bonding in Metal Carbonyls

M	$\chi^{(a)}$ [eV]	$\eta^{(a)}$ [eV]	ΔN	$\Delta H^{\ddagger}$ [kcal/mol][(b)]
V	2.24	1.24	0.211	Large[(c)]
Cr	2.47	1.58	0.192	46
Mn	3.10	2.19	0.149	37[(d)]
Fe	2.55	1.55	0.188	42
Co	4.12	3.04	0.091	22[(e)]
Ni	3.50	2.30	0.128	25
Cu	5.84	4.61	0.010	–
Mo	3.18	1.98	0.148	40
Ru	3.54	2.34	0.125	28
Pd	4.45	3.89	0.070	7[(f)]
Ag	6.87	5.57	−0.29	–
Pt	5.26	2.86	0.033	9[(f)]
Au	6.71	4.40	−0.24	–

[(a)] For low-spin valence state.
[(b)] Activation enthalpy for loss of first CO in known carbonyls. See References 4 and 5 for sources.
[(c)] $V(CO)_6$ is stable to dissociation.
[(d)] For $Mn_2(CO)_{10}$.
[(e)] For $Co_2(CO)_8$.
[(f)] Theoretical values (Reference 5).

The correlation between ΔN and $\Delta H^{\ddagger}$ is remarkable. The strongest bonds go with the largest values of ΔN. The cases where no stable carbonyls are known correspond to the smallest ΔN. The results for Cu, Ag and Au are consistent with the non-existence of $Cu_2(CO)_6$, $Ag_2(CO)_6$ and $Au_2(CO)_6$, the expected carbonyls. $Pd(CO)_4$ and $Pt(CO)_4$ do exist at low temperatures, but decompose at room temperature.

Bond energies and ΔN correlate in this case because the amount of π-donation is so important in determining the bond strength, and because the frontier orbitals are so nearly constant. If we keep the metal constant and vary the ligands, the results are not so good. For example, rating a number of common ligands in the order of decreasing π-bonding would give:

$$CS > CO \sim PF_3 > N_2 > PCl_3 > C_2H_4 > PR_3$$

$$\sim AsR_3 > R_2S > CH_3CN > \text{pyr} > NH_3 > R_2O$$

Various experimental criteria are used for this rating, especially IR spectra, but the order is conirmed by DFT calculations in a number of cases.[5] Bond energies also fall roughly in the same order, both from experiment and theory. However, the variation from one end to the other is quite small, about 10 kcal/mol. This results from the best σ-donors being at the end of the list, offsetting their poor π-accepting characters.

Taking a low-spin nickel atom as our metal, we next calculate ΔN for the ligands. The following is the order of decreasing ΔN:

$$CS > N_2 \sim PCl_3 > CO > PF_3 \gg \text{pyr} \sim CH_3CN > C_2H_4 \gg As(CH_3)_3$$

$$> P(CH_3)_3 \sim NH_3 > (CH_3)_2S > (CH_3)_2O$$

The last five compounds have ΔN negative, meaning they are net σ-donors.

The calculated ordering in ΔN is similar to that given by experiment, but there are discrepancies. These are due, for the most part, to orbital effects. For example, Figure 3.3(b) shows the π^* orbital of N_2, compared with that of CO in Figure 3.3(a). Clearly N_2 overlaps more poorly with a d orbital than CO does. Also the σ-orbital of CO, concentrated on carbon, is a better donor than that of N_2. An examination of the frontier MOs of pyridine and acetonitrile reveals that they also are poorly placed to give good overlap.[1]

Some variation of the orbitals can be tolerated, if the changes are due to substituents on a common substrate. In a series of olefins, the reactive orbitals are always the π and π^*, with perturbations due to the attached groups. An example of the Equation (3.1) would be the bonding of various olefins to Ni(0).[7] The data are the equilibrium constants in benzene for

$$NiL_3 + \text{olefin} \rightleftarrows NiL_2(\text{olefin}) + L \qquad K_{eq} \tag{3.6}$$

where L is a phosphite ligand.

Table 3.2 shows values of K_{eq} for various olefins, together with their χ and η values. Assuming that χ and η for NiL_2 are the same as for low-spin Ni, the values of ΔN have also been calculated by using Equation (3.1) and are shown in Table 3.2. These are for

$$NiL_2 + \text{olefin} \rightleftarrows NiL_2(\text{olefin}) \tag{3.7}$$

which is related to Equation (3.6) by a constant term.

The calculated ΔN values correlate very well with the equilibrium constants. Large positive values mean strong bonding, with π-bonding from metal to olefin dominating. Negative values mean that σ-bonding to the metal is greater than π-bonding. Clearly σ-bonding is less effective than π-bonding.

Ethylene has more π- than σ-bonding, in agreement with theoretical calculations for the reaction of $Ni(PH_3)_2$ with C_2H_4.[8] The calculated values of ΔN for the reactions of low-spin Pt and Pd are -0.0495 and $+0.005$, compared with that for Ni, $+0.053$. This agrees with the theoretical results for binding energies, $Ni > Pt > Pd$, especially if π-bonding is better than σ-bonding.

When we are primarily interested in the relative values of ΔN for a related series, approximate values of χ and η for the common reactant are usually adequate. But we can do a little better by referring back to Equation (2.29) of

Table 3.2 Electron Transfer in Reactions of Olefins with Low-Spin Nickel Atoms[a]

Olefin	K_{eq} [M]	χ [eV]	η [eV]	ΔN
Maleic anhydride	4×10^8	6.3	4.7	0.20
trans-NCCH=CHCN	1.6×10^8	6.2	5.6	0.17
CH_2=CHCN	4.0×10^4	5.4	5.6	0.12
C_2H_4	250	4.4	6.2	0.053
CH_2=CHF	90	4.2	6.1	0.042
Styrene	10	4.1	4.4	0.045
CH_3CH=CH_2	0.5	3.9	5.9	0.024
trans-2-Butene	2.7×10^{-3}	3.5	5.6	0.000
Cyclohexene	3.5×10^{-4}	3.4	5.5	−0.006
$(CH_3)_2C$=$CHCH_3$	3.0×10^{-4}	3.3	5.5	−0.013
$Ni(d^{10})$		3.5	2.3	
$Pd(d^{10})$		4.5	3.9	
$Pt(d^{10})$		5.3	2.9	

(a) Data from Reference 7.

Chapter 2. A version of this for the cases where we have unequal stoihiometries of the two reactants, as in the reaction[9]

$$M + nL = ML_n \tag{3.8}$$

gives the result

$$\mu = \frac{(\mu_M \sigma_M + n\mu_L \sigma_L)}{(\sigma_M + n\sigma_L)} = -\chi \tag{3.9}$$

where μ is the common value for ML_n after equilibrium is reached.

Thus we can calculate the change $(\mu - \mu_M)$ due to the attached ligands. Unfortunately, we know that Equation (3.9) is incomplete, since it gives only the initial effect. Further changes in μ will occur due to covalent and ionic bonding changes, which change ρ.[9] For example, take the reaction of two identical atoms:

$$\begin{array}{ccc} 2H & = & H_2 \\ -\mu = 7.2\,\text{eV} & & 6.7\,\text{eV} \end{array} \tag{3.10}$$

Here Equation (3.9) predicts no change.

Nevertheless, Equation (3.9) should give some idea of the effect of substituents on a reactive center. This is a problem of major importance in chemistry. Ziegler has used DFT to calculate the bond dissociation energies for a number of $Ru(CO_4)L$ complexes,[10]

$$Ru(CO)_4L(g) = Ru(CO)_4(g) + L(g) \qquad D_0 \tag{3.11}$$

Table 3.3 Values for RuL_4 Complexes Estimated from Equation (3.9)

	Ru	$Ru(CO)_4$	$Ru(PH_3)_4$	$Ru(NH_3)_4$
$-\mu$ [eV]	3.54	4.9	3.88	3.04

To estimate μ for $Ru(CO)_4$, we again start with a low-spin Ru atom and use Equation (3.9) to calculate a better value. The same may be done for substituents other than CO, such as the examples in Table 3.3. These results are very reasonable. By removing electron density from the metal atom, CO should make Ru more positive, and a poorer electron donor. Phosphine does not remove electron density so much as CO, and has a smaller effect. The result with NH_3 may be surprising at first, but it is also reasonable, since ammonia has no π-bonding tendency, and does give electron density to the metal by σ-bonding. The comparison between PH_3 and NH_3 is in agreement with theoretical calculations.[8,11] The π-bonding to PH_3 does not use d orbitals on P, but instead the anti-bonding σ^* orbitals of the P–H bond.

The predicted lower EN of $Ru(NH_3)_2$ compared with $Ru(PH_3)_2$ means that the former will be more reactive in cases where electron donation is required. An example would be the bonding of ethylene. The dissociation energy of $Ru(NH_3)_2C_2H_4$ is 64 kcal/mol, compared with 30 kcal/mol for $Ru(PH_3)_2C_2H_4$.[8] Of course $Ru(NH_3)_2$ would be much more difficult to prepare than $Ru(PH_3)_2$. Analogs of $Ru(PH_3)_2C_2H_4$ do exist, but not those of $Ru(NH_3)_2C_2H_4$.

In any case, there is a definite prediction about the effect of a neutral ligand on the reactivity of a transition metal. A ligand with a small value of χ will be activating for electron donation. This is almost the same as saying that a hard ligand will be activating. Hardness and small EN usually go together because the electron affinity has a large negative value.

There is ample evidence to support this prediction, since it simply says that hard ligands will favor a higher positive oxidation state for the central metal. We saw an example of this in Chapter 1, when the acidity of transition metal hydrides was discussed. It is also an example of the HSAB Principle, or the symbyiotic effect.

To obtain a value of χ adequate for more quantitative comparisons, we can also consider the further changes in μ that take place as the interacting fragments approach each other more closely. There is plenty of experimental evidence on this point, since we simply look at μ calculated from Equation (3.9), and the experimental value of μ for various ML_n.[9] In this case we need $L = CO$, and there are data for $Cr(CO)_6$ and $Fe(CO)_5$ (Table 3.4).[12]

It appears that covalent bonding leads to an increase in the electronic chemical potential.[9] As the interacting orbitals of two fragments get closer, the bonding ones go down in energy and the anti-bonding ones go up even more. Thus I is increased and A becomes more negative, so that $(I + A)$ gets smaller.

Table 3.4 Comparison of μ Values for ML_n Complexes

	$Cr(CO)_6$	$Fe(CO)_5$	$Ru(CO)_4$
$-\mu$(calc.) [eV]	5.4	5.5	4.9
$-\mu$(exp.) [eV]	3.9	4.4	(4.0) est.

With the estimated value of $\chi = 4.0\,\text{eV}$ for $Ru(CO)_4$, we can calculate ΔN for reaction (3.11) for a series of ligands, L. We still use $\eta = 2.3\,\text{eV}$, the value for low-spin Ru, since we do not know how to calculate the change in η due to the ligands in any easy manner. As we shall see later, η undoubtedly increases. Table 3.5 compares ΔN with D_0 for a series of paired ligands, where O is replaced by S.

We find that ΔN correlates with D_0, if we compare the paired ligands with each other: $CO < CS$, $CH_2O < CH_2S$; $CO_2 < CS_2$. Larger ΔN means stronger bonding, but not for $O_2 > S_2$, where the opposite occurs. The inversion for the last case is due to the promotion energy. The ground state for O_2 and S_2 is a triplet, whereas the valence state needed is a singlet with two electrons paired in a single π^* orbital. The pairing energy is 253 kJ/mol for O_2, but only 122 kJ/mol for S_2. This reduces the net bonding energy for oxygen.

Table 3.5 contains more information on the π-bonding in each case. There is an estimated value of the energy lowering, ΔE_π, and an estimate of the number of electrons transferred from the metal to the ligand, n_π. The values of ΔN and ΔE_π are correlated now in all cases. While ΔN and n_π are correlated, they are not equal. A major correction would be n_σ, the number of electrons donated from the ligand to the metal, but there are other factors as well.

D_0 values have also been calculated for CSe and CTe, as well as for CO and CS binding to $Ru(CO)_4$, and for Se_2 and Te_2 as well as O_2 and S_2.[10] The bonding for Se and Te is about the same as for S, and O is the one that differs. This is an illustration of a general phenomenon in chemistry. The behavior of the first-row

Table 3.5 Bonding in $Ru(CO)_4L$ Compounds[(a)]

L	χ [eV]	η [eV]	$D_0^{(b)}$ [kJ/mol]	ΔN	$n_\pi^{(c)}$	ΔE_π [kJ/mol]
CO	6.1	7.9	180	0.103	0.45	175
CS	5.8	5.6	237	0.130	0.55	216
CH_2O	5.0	5.9	181	0.061	0.74	346
CH_2S	4.9	4.4	228	0.067	0.76	360
CO_2	5.0	8.8	102	0.045	0.62	271
CS_2	5.35	5.56	158	0.085	0.83	318
O_2	6.26	5.82	133	0.140	1.04	594
S_2	5.51	3.85	244	0.122	0.91	443

(a) See Reference 10.
(b) D_0, dissociation energy.
(c) Number of electrons transferred from metal to ligand.

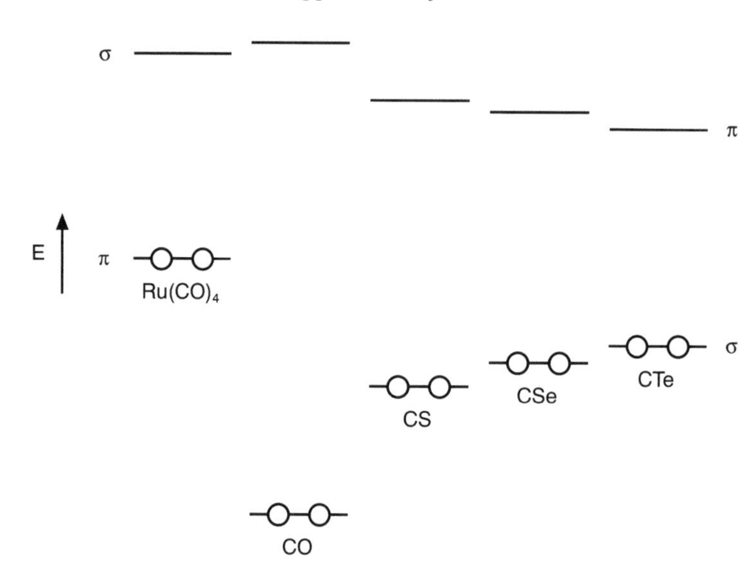

Figure 3.4 Frontier orbital energies on a relative scale for $Ru(CO)_4$, CO, CS, CSe and CTe.

elements (Li to F) is always different from that of the later rows, which do not differ much from each other.[13] This is particularly true for the non-metallic elements. Fluorine, with $I = 17.42\,\text{eV}$, differs markedly from chlorine, bromine and iodine, where $I = 12.97$, 11.81 and 10.45 eV. Also, the energy difference between ns and np is large for $n = 2$ and small for $n = 3$, 4, 5 and 6.

The result of these differences is illustrated in Figure 3.4, which shows the relative frontier orbital energies for CO, CS, CSe and CTe, compared with those of $Ru(CO)_4$. Carbon monoxide is much harder than any of the others. Therefore electron transfer, in both directions, will be much easier for CS, CSe and CTe. This leads to a larger value of D_0 in all cases. The stronger bonding of CS, compared with CO, and transition metals is well known.

These successful applications of Equation (3.1) to estimate bond energies are exceptions, rather than the rule. If ΔN is too large, so that ionic bonds are formed, then size factors will dominate the bonding. We want the bonding to be mainly due to electron transfer in one direction, but limited in extent. The best examples will be those where a coordinate covalent bond is formed. Charge transfer complexes should usually qualify, but only if similar molecules (and orbitals) are compared.[14]

RATES OF REACTION

Since ΔN is a measure of the initial interaction btween two particles, C and D, there is a better chance of finding a correlation between it and the rates of chemical reactions. That is, we have C–D as an

$$C + D \xrightarrow{k} C{-}D^{\ddagger} \rightarrow \text{product} \qquad (3.12)$$

activated complex, sitting on a potential energy maximum. The assumption is that large ΔN means a large lowering of he potential energy barrier to reaction, and a large value of k, the second-order rate constant. A restriction would be that we have an early transition state, so that $C{-}D^{\ddagger}$ resembles C and D more than it does the final product. Again, electron transfer mainly in one direction is needed.

A suitable example is the oxidative-addition reaction of transition metals and their complexes

$$ML_n + X{-}Y \rightarrow ML_nXY \quad \text{or} \quad ML_nX^+ + Y^- \qquad (3.13)$$

Take a ground-state iron atom as representative of ML_n. Then Table 3.6 shows the values of ΔN calculated from Equation (3.1) or the reaction with a number of common X–Y molecules. The results are in remarkable agreement with observations. Large positive values of ΔN are found for facile, fast reactions, and negative values for slow or unobservable reactions. Thse statements are based on studies of isolated iron atoms in low-temperature matrices, and on rate studies of various low-valent metal complexes reacting with these X–Y molecules.[16]

Reaction (3.12) proceeds by electrons being transferred from ML_n to an empty orbital which is anti-bonding for X and Y, usually the LUMO of X–Y.[2] The unreactive molecules in Table 3.6 have high-energy LUMOs, large negative

Table 3.6 Elecron Transfer in Reactions of X–Y with Ground-State Iron Atoms

X–Y	ΔN	X–Y	ΔN
F_2	0.26	HF	0.030
Cl_2	0.17	HCl	0.027
Br_2	0.15	H_2S	0.007
I_2	0.12	$C_2H{-}H$	0.004
H_2	0.11	$C_6H_5{-}H$	0.002
O_2	0.11	$CH_3{-}Cl$	−0.013
$CCl_3{-}Cl$	0.10	$C_6H_5O{-}H$	−0.015
$NO_3{-}H$	0.094	$CH_3{-}F$	−0.035
HI	0.072	H_2O	−0.036
H–CN	0.066	$CH_3{-}H$	−0.060
$CH_3{-}I$	0.048	$NH_2{-}H$	−0.065
$SiH_3{-}H$	0.035	$CH_3O{-}H$	−0.066

values for A, and small values for χ. Hence ΔN is negative, indicating a positive contribution to the activation energy barrier, making it higher.

Naturally, other factors that influence rates of reaction cannot be ignored. For example, the rate of addition of HX molecules is some 10^7 times faster than the addition of CH_3X. This refers to the concerted addition of molecular HX to $IrCl(CO)[P(C_6H_5)_3]_2$ in solvents such as toluene.[17] Clearly steric hindrance is much greater in CH_3X than in HX and accounts for the slower reaction.

In agreement with Table 3.6, oxidative addition of SiH_4 is much more facile than that of CH_4. An *ab-initio* MO calculation has been made of the reactions of both SiH_4 and CH_4 with $Pt(PH_3)_2$.[18] The activation barrier for CH_4 is calculated to be 28 kcal/mol, and there is no barrier for SiH_4. As expected, there is an early TS for SiH_4 and a late one for CH_4. Experimentally it is found that phenols react readily with $Pd(PR_3)_3$ in toluene, and alcohols do not react.[19] This is in agreement with ΔN being much more negative for CH_3OH than for C_6H_5OH in Table 3.6.

In addition to changing the substrate, we can also change the metal atom. Metals of lower electronegativity than iron should be more reactive, and metals of higher χ should be less reactive, up to a point. If χ for the metal becomes comparable with χ for the atoms or radicals X and Y, then electron transfer in one direction is not required, and Equation (3.1) is no longer a good criterion. There is limited evidence available on free transition metal atoms in low-temperature matrices. The most reactive atoms are Sc, Ti and V, as expected.[16] As we shall see in Chapter 5, there is reason to believe that there is a parallelism between the reactivity of isolated metal atoms and the atoms in the bulk metal. Thus the free atoms of the noble metals, such as Os, Ir, Pt and Au, are expected to be slow to react.

But there is an opposing effect. The strength of the bonds to the noble metals are usually the strongest, when comparing the metals in a given triad. The order most often found is $5d \sim 3d \gg 4d$. Particularly when the bond to the metal is a pure σ-bond, as is the case for H or CH_3, the most EN metals form the strongest bonds.[20] When π-bonding is dominant, as in the metal carbonyls, the situation is reversed.

There is a great deal of information available on rates of oxidative addition for metal complexes,[21] but we can only make meaningful comparisons between metals if the ligands and oxidation state of the metal are held constant. This again means comparing metals in a triad. The same result is found for the rates of reaction as for the bond strengths, $5d \sim 3d \gg 4d$. This suggests that for reaction (3.12) we usually have a late transition state resembling the products. The final energy, including the bond strengths of M–X and M–Y, determines the height of the energy barrier.

A clear prediction is made about the effect of the auxiliary ligands, L, on the ease of oxidative addition for a given M. Ligands with a small value of χ will increase the reactivity. That is, hard ligands will make the complex a better reducing agent, or a better electron donor. Equally we can say that hard ligands will increase the strength of M–X and M–Y bonds.

A good example of this predicted effect is given by *ab-initio* MO calculations on the systems $NiLH_2$ and $PdLH_2$.[22] The reaction studied is

$$ML + H_2 = MLH_2 \qquad \Delta E \tag{3.14}$$

so that a negative ΔE means MLH_2 is stable. The results found for neutral ligands are listed in Table 3.7. The large stabilizing effect of the hard ligand, H_2O, stands out. On the other hand, the bond energy between Ni and H_2O, is very low, 5 kcal/mol, compared to 14 to 40 kcal/mol for the other ligands.

While Equation (3.1) predicts that H_2O will donate electron density to the nickel atom, this is not what actually happens at the bonding distance of 2.33 Å. The bonding is van der Waals in character, being half dipole–induced dipole and half correlation energy.[23] There is little, if any, electron transfer in either direction. This result is not a violation of DFT since the chemical potential of $Ni-H_2O$ is made constant everywhere by the changes in electron density which do occur.

The weak bonding of H_2O and NH_3 to neutral atoms means that it is not easy to make complexes conaining both a metal atom in a low oxidation state and hard ligands. An attempt to make $Rh(NH_3)_4^+$ in aqueous solution would certainly lead to the formation of $Rh(NH_3)_4HOH^+$ instead. The Rh(III) hydride would result from the oxidative cleavage of water.[24] A Rh(I) complex containing four nitrogen donor atoms has been made, $Rh(C_2DOBF_2)$, where C_2DOBF_2 is a complicated tetradentate ligand.[25] It is very reactive towards oxidative addition.

In a more straightforward example, a detailed study has been made of the reaction

$$IrX(CO)[P(C_6H_5)_3]_2 + CH_3I \rightarrow CH_3IrIX(CO)[P(C_6H_5)_3]_2 \tag{3.15}$$

The order of rates found for different X^- was $F^- > N_3^- > CL^- > Br^- > NCO^- > I^- > NCS^-$, with hard F^- reacting 100 times faster than soft (S-bonded)

Table 3.7 Energy Changes for the H_2 Addition to ML

	ΔE [kcal/mol]	
L	Ni	Pd
None	−6	−9
C_2F_4	+9	–
C_2H_4	+2	+13
CO	+3	+25
N_2	−9	+3
PH_3	−9	+4
H_2O	−17	−19

NCS^-.[26] Results like these, and the others cited above, are rather unexpected in more classical thinking. Soft ligands were thought to put negative charge on the metal atom, making it a better electron donor.

Ethylene, with $\chi = 4.4$ eV and $\eta = 6.2$ eV, is a typical organic molecule in being intermediate in EN so that reaction with both electrophiles and nucleophiles is possible. For substituted olefins, χ ranges from 3.0 eV for $(CH_3)_2=C(CH_3)_2$ to 7.3 eV for $(NC)_2C=C(CN)_2$. More EN olefins react best with electron donors, or nucleophiles, as we have already seen in Table 3.2. The least EN olefins react best with reagents like Br_2 and H_3O^+. In the case of Br_2, with $\chi = 6.6$ eV, there is good agreement between the rate constants and ΔN calculated from Equation (3.1), as shown in Table 3.8.[27]

In the Diels–Alder reaction, butadiene ($\chi = 4.3$ eV) is less reactive towards olefins than cyclopentadiene ($\chi = 3.8$ eV). This suggests that the diene is the electron donor to the olefin

$$C_5H_6 + C_2H_4 \rightarrow C_7H_{10} \tag{3.16}$$

In that case the most EN olefins should react the fastest. Table 3.9 shows that this is the case, with near-perfect agreement between the ordering of ΔN and k, the second-order rate constant. The very low reactivity of ethylene itself is not explained, however. Cyclopentadiene reacting with itself as an olefin is included in the table, and it may be the one in the table that is out of line.

An unusual application of Equation (3.1) was made in a study of anionic polymerization of olefins as a function of the olefin and of the initiator, AB.[28]

Table 3.8 Values of ΔN and Relative Rates of Olefin Reactions with Bromine in Methanol

Reactant	χ [eV]	η [eV]	ΔN	$k_{rel}^{(a)}$
$CH_2{=}CH_2$	4.4	6.2	0.11	1
$CH_2{=}CHCH{=}CH_2$	4.3	4.9	0.13	$\sim 50^{(b)}$
$CH_2CH{=}CH_2$	3.9	5.9	0.14	61
trans-$CH_3CH{=}CHCH_3$	3.5	5.6	0.16	1700
cis-$CH_3CH{=}CHCH_3$	3.45	5.7	0.16	2000
$(CH_3)_2C{=}CH_2$	3.5	5.7	0.16	5400
$(CH_3)_2C{=}CHCH_3$	3.3	5.5	0.17	1.3×10^5
$(CH_3)_2C{=}C(CH_3)_2$	3.0	5.3	0.19	1.8×10^6
$CH{\equiv}CH$	4.4	7.0	0.10	10^{-3}
$CH_2{=}CHCN$	5.4	5.6	0.06	V. slow
$CH_2{=}CHCHO$	$5.3^{(a)}$	$4.9^{(a)}$	0.08	V. slow
$CH_2{=}C{=}CH_2$	3.8	5.1	0.14	$\sim 12^{(b)}$
$CH_2{=}CHOAc$	$4.3^{(a)}$	$5.5^{(a)}$	0.12	$\sim 120^{(b)}$

[a] After Reference 27.
[b] Estimated from rates of hydration.

Table 3.9 Calculated values of ΔN and Rate Constants for Reactions of Olefins with 1,3-cyclopentadiene

Reactant	χ [eV]	η [eV]	ΔN	k [$M^{-1}\,s^{-1}$][(a)]
$C_2(CN)_4$	7.3	4.5	0.19	4.3×10^8
$NCCH{=}C(CN)_2$	6.8	4.7	0.17	4.8×10^6
$CH_2{=}C(CN)_2$	6.5	4.9	0.14	4.6×10^5
Maleic anhydride	6.3	4.7	0.13	5.5×10^4
p-Benzoquinone	5.7	3.9	0.11	9.0×10^3
Maleonitrile	6.2	5.6	0.12	9.1×10^2
Fumaronitrile	6.2	5.6	0.12	8.1×10^2
$CH_2{=}CHCN$	5.4	5.6	0.08	10
C_2H_4	4.4	6.2	0.03	10^{-4}[(b)]
Cyclopentadiene	3.8	5.8	0.00	0.9

[(a)] At 20°C in dioxane. See Reference 29.
[(b)] Estimated from gas-phase data.

The mechanism of such polymerizations is believed to start with atack of an anion B^-, from the initiator, on the olefin (Equation (3.17)). The new anion thus formed attacks the monomer in the chain propagation step (Equation (3.18)).

$$AB + CH_2{=}CHR \rightarrow B{-}CH_2{-}CHR^- + A^+ \tag{3.17}$$

$$BCH_2{-}CHR^- + CH_2{=}CHR \rightarrow B(CH_2CHR)_2^- \text{ etc.} \tag{3.18}$$

The role of A^+ and the terminating steps are not well understood. Since the reactions are run in solvents such as dioxane or tetrahydrofuran, ion-pairing and aggregation are occurring.

Eleven olefins and ten initiators were considered in all. The initiators were $NaCH_3$, $LiCH_3$, n-C_4H_9Li, t-C_4H_9Li, CH_3MgCl, $NaOCH_3$, $LiOCH_3$, C_5H_5N, $(CH_3)_3N$ and H_2O, in order of decreasing efficiency. Since there were no experimental values of I or A available for the first eight of these, they were all calculated by *ab-initio* MO methods. The energies of AB, AB^+ and AB^- were calculated separately to find I and A for all the initiators. To have consistency, the same was done for the 11 olefins. Comparison of the χ values found in this way with those that were known from experiment showed that the right ordering was obtained, though the theoretical values were about 0.8 eV lower, except for $(CH_3)_3N$ and H_2O.

The values of ΔN were calculated from Equation (3.1) for all 110 possible combinations. Table 3.10 gives some representative results. The olefins are in order of increasing reactivity, as found experimenally, with the most reactive at the bottom. The values of ΔN calculated for all ten initiators put all the olefins in the right order of reactivity, the largest value of ΔN being for $CH_2{=}C(CN)_2$.

Table 3.10 Values of $2\Delta N$ Calculated for Initiation Reaction[a]

Monomer	Initiator				
	CH_3Na	n-BuLi	CH_3MgCl	CH_3ONa	C_4H_5Ne
Butadiene	0.030	0.005	−0.109	−0.042	−0.024
Methyl crotonate	0.122	0.096	−0.031	0.049	0.046
Methyl methacrylate	0.122	0.096	−0.031	0.049	0.046
Crotononitrile	0.144	0.118	−0.011	0.072	0.064
Methacrylonitrile	0.147	0.121	−0.007	0.075	0.066
Methyl acrylate	0.156	0.130	0.000	0.084	0.073
Acrylonitrile	0.180	0.154	0.021	0.108	0.092
Methyl vinyl ketone	0.185	0.158	0.022	0.111	0.094
Methyl α-cyanoacrylate	0.253	0.223	0.071	0.173	0.142
Niroethylene	0.257	0.229	0.082	0.181	0.149
Vinylidene cyanide	0.307	0.276	0.116	0.226	0.183

[a] After Reference 28.

Placing the initiators in the right order was not successful. ΔN should have decreased, or become negative, going from left to right in the table for any one monomer. This is only roughly true, CH_3MgCl having ΔN too small, and all the molecular initiators having ΔN too large. Considering the complex nature of a Grignard solution, and the fact that molecular initiators must have different mechanism, these failures are not surprising. The success with the monomer ordering shows that the transfer of negative charge to the olefin must always play a key role.

REACTIONS OF FREE RADICALS

A free radical, or univalent atom, is a chemical system like any other: χ and η can be found for it, and Table 3.11 shows a listing of such data for a number of important radicals. The acid–base character of free radicals has been recognized for some time.[30] It is common to speak of electrophilic radicals, such as Cl, and nucleophilic radicals, such as $(CH_3)C$. Table 3.11 is a quantitative ordering of these descriptions. The alkali metal atoms could also be added to the list. These would be the most nucleophilic, or best electron donors.

Table 3.12 lists the rate constants for a series of radicals reacting with ethylene. The calculated values of ΔN are also listed and are nearly in the same order. H and OH react somewhat faster than expected. The more electrophilic radicals

Table 3.11 Experimental Values for Radicals[(a)]

Radical	I [eV]	A [eV]	χ [eV]	η [eV]
F	17.42	3.40	10.41	7.01
OH	13.17	1.83	7.50	5.67
NH_2	11.40	0.78	6.09	5.31
CH_3	9.82	0.08	4.96	4.87
Cl	13.01	3.62	8.31	4.70
SH	10.41	2.30	6.40	4.10
PH_2	9.83	1.25	5.54	4.29
SiH_3	8.14	1.41	4.78	3.37
Br	11.84	3.36	7.60	4.24
SeH	9.80	2.20	6.00	3.80
I	10.45	3.06	6.76	3.70
H	13.59	0.74	7.17	6.42
HO_2	11.53	1.19	6.36	5.17
CN	14.02	3.82	8.92	5.10
NO_2	11.25[(b)]	2.95	7.10	4.15
NCO	11.76	3.6	7.68	4.08
$Si(CH_3)_3$	6.5	1.0	3.75	2.75
OCH_3	8.6	1.57	5.10	3.50
CH_2CN	10.0	1.54	5.77	4.23
SC_6H_5	8.63	2.47	5.50	3.08
OC_6H_5	8.85	2.35	5.60	3.25
C_2H_5	8.38	−0.39	4.00	4.39
$i\text{-}C_3H_7$	7.57	−0.48	3.55	4.03
$t\text{-}C_4H_9$	6.93	−0.30	3.31	3.61
C_6H_5	8.95	0.10	5.20	4.10
C_2H_3	8.95	0.74	4.85	4.10
CHO	9.90	0.17	5.04	4.88
$COCH_3$	8.05	0.40	4.23	3.82
$CH_2C_6H_5$	7.63	0.88	4.26	3.38
CCl_3	8.78	2.35	5.57	3.23
CH_3	9.25	1.84	5.55	3.71
$SiCl_3$	7.92	2.50	5.20	2.70
NO	9.25	0.02	4.63	4.61
CH_3S	8.06	1.86	4.96	3.10
$GeCl_3$	8.5	2.8	5.65	2.85
$Sn(CH_3)_3$	7.10	1.7	4.40	2.70

[(a)] Data from reference 32.
[(b)] Vertical value. O. Edquist, E. Lindholm, L.E. Selin, H. Sjogren and L. Åstorink, Phys. Scr., *1*, 172 (1970)

react the fastest. Electrons are being removed from the π orbital of ethylene, converting the double bond to a single one:

$$R\cdot + C_2H_4 \xrightarrow{k} R{-}CH_2{-}CH_2\cdot \tag{3.19}$$

Table 3.12 Values of ΔN for Reaction of Free Radicals with Ethylene[a]

Radical	χ [eV]	η [eV]	ΔN	k [M^{-1} s^{-1}]
Cl	8.3	4.7	0.179	4.5×10^{10}
NCO	7.7	4.1	0.160	1.1×10^{9}
OH	7.5	5.7	0.131	5.0×10^{9}
H	7.2	6.4	0.111	2.0×10^{9}
Br	7.6	4.2	0.154	1.0×10^{8}
CF_3	5.5	3.7	0.056	3.5×10^{6}
O_2H	6.4	5.2	0.086	2.0×10^{6}
CCl_3	5.4	3.0	0.049	4.5×10^{4}
CH_3	5.0	4.9	0.027	4.5×10^{4}
C_2H_5	4.0	4.4	−0.019	3.5×10^{4}
i-C_3H_7	3.6	4.0	−0.042	2.2×10^{4}
t-C_4H_9	3.3	3.6	−0.056	8.9×10^{3}

[a] Gas-phase reactions at 437 K. Data as given in Reference 27, except for NCO, Reference 31.

We can reverse the order of reactivity of the radicals by going to a reactant which is more electronegative than C_2H_4. A good example is O_2, to which most radicals add, with $\chi = 6.3\,\text{eV}$

$$R + O_2 \xrightarrow{k} R{-}O_2\cdot \tag{3.20}$$

Table 3.13 compares ΔN with the experimental second-order rate constant for a number of radicals. Certainly the most nucleophilic radicals now react the fastest, though the ordering is not perfect. The phenoxide and methoxide radicals react more slowly than predicted.

To understand this, recall that only an early transition state should show a correlation between ΔN and k. A late TS is more likely to correlate k with the exothermicity of the reaction, or the $R{-}O_2$ bond strength.[34] Since O–O bonds are notoriously weak, a slow reaction is not unexpected for RO· radicals. It is likely that the same explanation serves for the non-reactivity of CH_3S, and part of the inertness of HS, though we usually think of the S–O bond as strong.

This S–O bond, however, is not strong since addition of RS· to O_2 will lead to the formation of a derivative of sulfenic acid, RSOR′. Such derivatives are very unstable. Stable S–O bonds are the ones present in the isomeric sulfoxides, RR′SO, which are coordinate covalent bonds. In HSAB terms, RSOR′ is a combination of a soft acid, RS^+, with a hard base, $R'O^-$, while sulfoxide is a combination of a soft acid, oxene (the ^1D oxygen atom), with a soft base, RSR′.

Table 3.13 Rate Constants for the Reaction of Radicals with O_2 in the Gas Phase[a]

Radical	ΔN	$k \times 10^{12}$ [a] $[cm^{-3}/mol\,s]$[b]
$t\text{-}C_4H_9$	0.158	23.4
$i\text{-}C_3H_7$	0.135	14.1
C_2H_5	0.112	4.4
$CH_2C_6H_5$	0.108	12.0
C_2H_3	0.094	6.7
SiH_3	0.082	13.0[c]
SCH_3	0.072	$>2 \times 10^{-5}$ [d]
OCH_3	0.064	2×10^{-3}
CH_3	0.062	2.0
CCl_3	0.038	1.4[e]
OC_6H_5	0.038	$<10^{-2}$
CH_2CN	0.025	1×10^{-2} [f]
SH	−0.005	$<10^{-5}$

[a] Data rom Reference 33, except where otherwise indicated.
[b] At 298 K.
[c] Reference 35.
[d] Reference 36.
[e] Reference 37.
[f] Reference 38.

Table 3.14 Values of ΔN and Rate Constants for Reaction of OH Radical with Olefins[a]

Olefin	χ [eV]	η [eV]	ΔN	$k \times 10^{12}$ $[cm^3/mol\,s]$[b]
$(CH_3)_2C{=}C(CH_3)_2$	3.0	5.3	0.206	103
$(CH_3)_2C{=}CHCH_3$	3.3	5.5	0.186	85
$CH_3CH{=}CHCH_3$[c]	3.5	5.6	0.177	42
$(CH_3)_2{=}CH_2$	3.5	5.7	0.175	55
$CH_3CH{=}CH_2$	3.9	5.9	0.155	25
$CH_2{=}CHCH{=}CH_2$	4.3	4.9	0.151	70
cis-$CHCl{=}CHCl$	4.3	5.4	0.145	2.4
trans-$CHCl{=}CHCl$	4.4	5.2	0.142	2.1
$CCl_2{=}CCl_2$	4.5	4.8	0.143	0.2
$CG_2{=}CHCl$	4.4	5.6	0.137	7
$CH_2{=}CCl_2$	4.6	5.3	0.134	10
$CH_2{=}CH_2$	4.4	6.2	0.130	8.5

[a] Data from Reference 39.
[b] At 298 K, gas phase.
[c] *Cis* and *trans* both the same.

Table 3.15 Comparison of χ_P and χ_A

	χ_P	χ_A
t-C_4H_9	2.82	3.31
i-C_3H_7	2.78	3.55
C_2H_5	2.77	4.00
C_2H_3	2.77	4.85
CH_3	2.76	4.96
SH	2.61	6.40
H	2.30	7.17

Another variation of Table 3.12 is to change the olefin, while keeping the radical constant. Table 3.14 shows some results for the reaction with OH radical, $\chi = 7.5\,\text{eV}$ and $\eta = 5.7\,\text{eV}$. As expected, the least electrophilic olefins react the fastest. As the large values of ΔN suggest, there is no energy barrier for most of these reactions. Instead, electron transfer must be increasing the frequency of collision, or the duration of the collision. Only tetrachloroethylene has an appreciable energy barrier of 2.6 kcal/mol. This shows up as a reduction in the rate due to steric hindrance.

It is worth nothing that the use of both I and A for both reactants gives the best correlation with the rate constants. The more common procedure in frontier orbital theory is to consider only I for the donor and A for the acceptor.[40] This works for the alkyl olefins, but fails badly for the chloro olefins. If a nuleophilic radical, such as $(CH_3)_3C$, is used as the common reagent, then the order in Table 3.14 is completely inverted.[41] The chloro olefins now react very rapidly, and the alkyl olefins are slow.

Pauling electronegativities (χ_p) are also available for many radicals or groups. These are obtained, not from the free radicals, but from properties of molecules containing the groups. Various properties have been used, and this leads to some fluctuations in the values calculated. Some typical results[42] (Table 3.15) may be compared with the absolute values (χ_A) of Table 3.11. While the units are not the same, the fact that the orders are reversed suggests correctly that the Pauling ENs will not be useful in discussing free-radical reactions.

AROMATIC ELECTROPHILIC SUBSTITUTION

The chemistry of aromatic compounds was one of the early testing grounds for the application of quantum mechanics to chemical problems. The reason for this was chiefly the simplicity and success of Hückel molecular orbital (HMO) theory. It is appropriate to see how well DFT explains aromatic behavior. We already have one example of this in Chapter 2: aromatic stability can be correlated with $(I - A)$, chemical hardness.

The most characteristic, and well studied, reaction of aromatic compounds is electrophilic substitution, such as nitration. There is general agreement on the detailed mechanism; for example,[43]

$$\text{C}_6\text{H}_6 + \text{E}^+ \xrightarrow{k} \underset{\text{C}}{\text{C}_6\text{H}_6\text{E}^+} \xrightarrow{\text{fast}} \text{C}_6\text{H}_5\text{E} + \text{H}^+ \tag{3.21}$$

The rate-determining step is the formation of a σ-complex, C, also called the Wheland intermediate.[44] This is followed by the rapid loss of a proton. The activated complex for the formation of C is thought to resemble C very closely. The possible formation of a π-complex, prior to C, is usually of little consequence.

Taking benzene as the prototype, its value of $\chi = 4.1\,\text{eV}$ shows that it can be an electron donor, or base. It will form acid–base complexes with molecules well above it in χ (see Table 2.2 of Chapter 2), such as Cl_2, SO_3, BF_3 and HNO_3. Since $\eta = 5.3\,\text{eV}$ is relatively soft, it will also form complexes with soft metal ions, such as Ag^+, as well as hard ions, such H^+, which are sufficiently strong.

Figure 3.5 shows the orbital energies for Cl_2, C_6H_6 and $C_6H_5NO_2$. Electrons will flow spontaneously from benzene to chlorine. Nitrobenzene will not donate

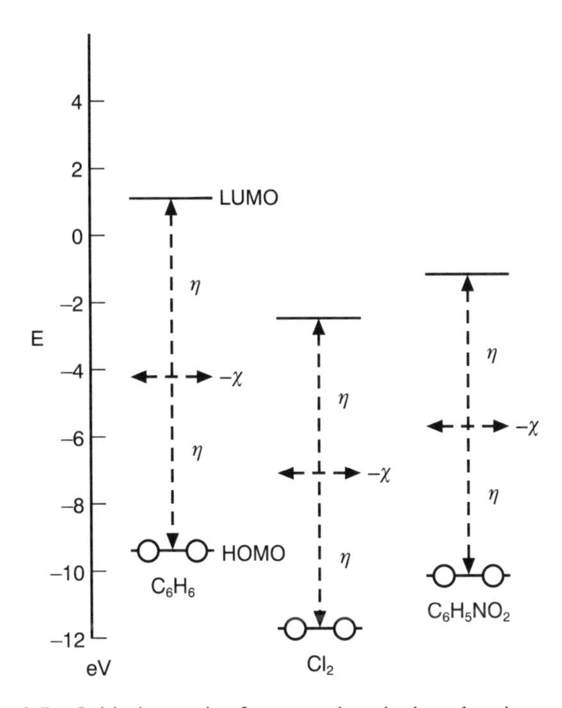

Figure 3.5 Orbital energies for several molcules, showing χ and η.

electrons to chlorine quite as readily. Taking Cl_2 as a common reagent, we can calculate ΔN for the substituted benzenes where I and A are known. The results are given in Table 3.16.

Making the usual assumption that a large value of ΔN means a large lowering of the activation energy, the results are excellent since a near-perfect order of reactivity towards electrophilic substitution is generated.[45] Some heteroaromatic molecules are also listed in Table 3.16. The ΔN values again vary just like the relative reactivities (Table 3.17).[46]

Also included in Table 3.16 are some polynuclear aromatics. They also are in the right order, with anthracene being the most reactive and benzene the least.

Table 3.16 Values of ΔN Calculated for Reaction of Cl_2 with Aromatic Compounds

Reactant	χ [eV][(a)]	η [eV][(a)]	ΔN
$C_6H_5NHCH_3$	3.05	4.25	0.22
$C_6H_5N(CH_3)_2$	3.10	4.35	0.22
$C_6H_5NH_2$	3.3	4.4	0.21
$C_6H_5OCH_3$	3.55	4.65	0.19
1,3,5-$C_6H_3(CH_3)_3$	3.7	4.7	0.18
p-$C_6H_4(CH_3)_2$	3.7	4.8	0.18
C_6H_5SH	3.8	4.6	0.17
C_6H_5OH	3.8	4.8	0.17
$C_6H_5CH_3$	3.9	5.0	0.16
$C_6H_5CH{=}CH_2$	4.1	4.4	0.16
C_6H_5I	4.1	4.6	0.16
C_6H_5Br	4.1	4.8	0.15
C_6H_5Cl	4.1	4.9	0.15
C_6H_5F	4.1	5.0	0.15
C_6H_6	4.1	5.3	0.14
$C_6H_5CO_2CH_3$	4.7	4.6	0.12
$C_6H_5COCH_3$	4.8	4.5	0.12
C_6H_5CHO	5.0	4.6	0.11
$C_6H_5CO_2H$	4.9	4.8	0.11
C_6H_5CN	5.0	4.7	0.11
$C_6H_5NO_2$	5.5	4.4	0.08
p-$C_6H_4(NO_2)CN$	6.1	4.5	0.05
Thiophene	3.8	5.0	0.17
Furan	3.5	5.3	0.18
Pyrrole	2.9	5.4	0.21
Anthracene	3.8	3.3	0.26
Azulene	4.1	3.3	0.18
Phenanthrene	4.1	3.8	0.17
Naphthalene	4.0	4.2	0.17
Biphenyl	4.0	4.3	0.17

[(a)] Reference 27.

Table 3.17 Values of ΔN and k_{rel} for Reaction of Cl_2 with Benzene and Heteroaromatic Compounds

	Benzene	Thiophene	Furan	Pyrrole
ΔN	0.14	0.17	0.18	0.21
k_{rel}	1	10^3	10^5	10^{12}

Other condensed aromatic ring compounds will also fall in line, since as the softness increases, both ΔN and reactivity will increase. We also see that multiple substitution by NO_2 or CN groups will increase χ. This will make nucleophilic substitution possible, if a suitable leaving group is also present.

None of the above results addresses the question of the position of reaction in the aromatic molecule. In DFT this is done by considering the Fukui function, f. It is interesting that this orientation problem was also the topic of the first paper on frontier orbital theory.[47] Reaction was predicted to occur at the position of highest frontier orbital (FO) electron density. The frontier orbital in electrophilic substitution would be the HOMO. If this orbital were written as the usual linear combination of atomic orbitals, then the density at each atom would simply be the square of the coefficient in the LCAO, or c_i^2 where i indicates the atom. Since this is also one of the ways of approximating f, the success of the FO method may also be claimed for DFT. However, the details of the Fukui function application will be postponed briefly to look at a method unique to density functional theory, and using the concept of hardness.[48]

We can see from Equation (3.20) that the intermediate C is a different π-system from the reactant benzene. In terms of the π-electrons only, it is a linear pentadienyl cation, $C_5H_5^+$. Thus there is a change in the π-electron energy. Let us assume that the activation energy for electrophilic substitution comes only from this π-energy change. We can readily calculate this from simple HMO theory. Figure 3.6 shows the resulting orbital energies in terms of α and β, the usual coulomb and exchange integrals.

The energy of the π-electrons of C_6H_6 is $6\alpha + 8\beta$, and that of $C_5H_5^+$ is $4\alpha + 5.46\beta$. The change of 2α is added to the energy change in the σ-system, which is then ignored by assuming it to be constant for various similar molecules. The energy change of -2.54β was called the cation localization energy, L^+, by Wheland.[44] Changes in L^+ from one aromatic system to the next were then responsible for their differing reactivities.

This assumption was very successful when applied to the problem of the orientation of substitution in a molecule where there was more than one kind of carbon atom. For example, naphthalene can react at either the 1- or 2-position. Reaction occurs primarily at the 1-position, and L_1^+ is indeed smaller than L_2^+. The use of L^+ was not always successful for changes in the reacting molecule, presumably because differences in the σ-energies became important.

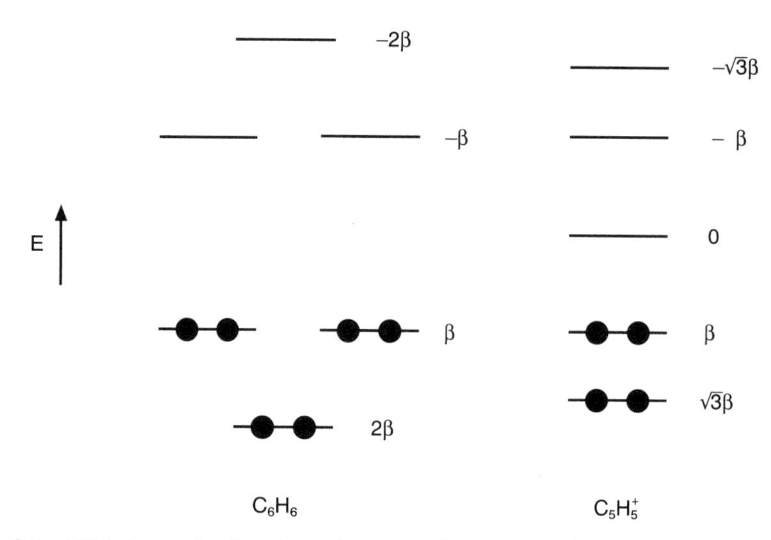

Figure 3.6 Hückel energies for the π-electrons of benzene and the pentadienyl cation. The zero of energy is α.

Figure 3.6 can also be used to calculate μ and η for the reactant and the intermediate. The orbital energies in HMO theory are given by

$$\varepsilon_i = \alpha + X_i\beta \tag{3.22}$$

where the X_i are numbers which are fixed for the particular conjugated molecule, once the connectedness is given. From the definitions of μ and η, we have

$$\mu = \alpha + \beta(X_{\mathrm{LUMO}} + X_{\mathrm{HOMO}})/2 \tag{3.23}$$

$$\eta = \beta(X_{\mathrm{LUMO}} - X_{\mathrm{HOMO}})/2 \tag{3.24}$$

For C_6H_6 we have $\mu = \alpha$ and $\eta = -\beta$, remembering that both α and β are negative. For the intermediae C, $\mu = (\alpha + \beta/2)$ and $\eta = -\beta/2$.

The question to be answered is whether changes in η, or μ, on going from reactant to transition state, offer any clue as to the magnitude of the activation energy. The complex C is assumed to be close in energy to the transition state. If we define the activation hardness as

$$\Delta\eta^{\ddagger} = (\eta_{\mathrm{R}} - \eta_{\mathrm{TS}}) \tag{3.25}$$

then in the case at hand, $\Delta\eta^{\ddagger}$ is -0.50β. Similar calculations can be made for electrophilic substitution at any of the positions in other aromatic molecules.

Table 3.18 Reactivity Indices for Substituted Benzenes

Compd	Position	$\Delta\eta^{\ddagger}$ [a]	Observed product (% per site) [b,c]
C_6H_5F	2	0.462	6
	3	0.492	0.5
	4	0.435	87
C_6H_5Cl	2	0.480	15
	3	0.494	0
	4	0.462	70
C_6H_5Br	2	0.483	19
	3	0.494	0
	4	0.463	62
C_6H_5OH	2	0.421	20
	3	0.486	0
	4	0.363	60
$C_6H_5NH_2$	2	0.391	
	3	0.484	*Ortho*, *para* directing
	4	0.307	
$C_6H_5CH_3$	2	0.392	28.5
	3	0.485	1.5
	4	0.339	40
C_6H_5CHO	2	0.269	9.5
	3	0.139	36
	4	0.276	9
$C_6H_5CO_2H$	2	0.322	9.3
	3	0.222	40.1
	4	0.325	1.3

[a] In units of $-\beta$.
[b] For nitration.
[c] For references, see Reference 48.

What Zhou and Parr found was that the smaller the activation hardness is, the faster is the reaction.[48] Thus $\Delta\eta^{\ddagger}$ is a reactivity index. Table 3.18 shows the results for the amounts of *ortho-*, *para-* and *meta-*substitution in the nitration of substituted benzenes. Similar good results were found for the site selectivity in a large number of condensed-ring hydrocarbons and heterocyclic molecules.

It can be seen in Table 3.18 that there is no correlation between reactivity and the activation hardness, if the molecule is changed. Thus benzoic acid has a smaller $\Delta\eta^{\ddagger}$ than benzene, but is much less reactive. Again, changes in the σ-bonding have become important, similarly to the case of the cation localization energy. Actually both L^+ and $\Delta\eta^{\ddagger}$ do correlate with reactivity, if only the condensed-ring hydrocarbons are compared with each other.[48]

It is difficult to give a proof as to what the exact relationship between the activation hardness and the energy barrier should be, or even that a relationship

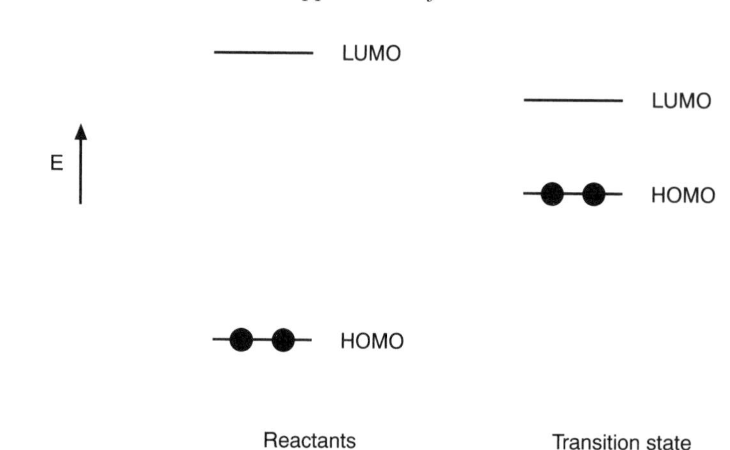

Figure 3.7 Typical changes in hardness (HOMO–LUMO gap) for a chemical reaction. The harder the TS, the more stable it is, and the easier it is to reach. After Reference 48.

should exist. Figure 3.7, however, shows that there are good reasons to believe that a small value of $\Delta\eta^{\ddagger}$ should correspond to a low activation energy. The transition state is shown to be softer than the reactant in Figure 3.7. This follows from the general properties of the activated complex, which is on a maximum in the potential energy–reaction coordinate plot. Therefore it decomposes immediately upon formation.

From the perturbation theory of chemical reaction, this means that there must be low-lying excited states for the activated complex, or that the HOMO–LUMO gap is small.[49] From the same theory we can assume that the softer the transition state, the more unstable it is, and the more difficult it will be to form. A hard TS will be more stable, and hence easier to form (low energy barrier). We can see in Figure 3.7 that increasing the hardness of the TS will mean lowering the HOMO of the activated complex, and reducing the energy difference between it and the reactant.

FUKUI FUNCTIONS AND ATOMIC CHARGES

The natural DFT quantity to use in probing site selectivity within a molecule is the Fukui function, $f(r)$. The definition is

$$f(r) = (\partial\mu/\partial v)_N = (\partial\rho(r)/\partial N)_v \tag{3.26}$$

where $f(r)$ is used to remind us that it is a function of position, as is $\rho(r)$. It is necessary to approximate $f(r)$ in most cases. Taking the case of $f^-(r)$, for electrophilic attack on the molecule, we have several approximations. For example,

$$f^-(r) = [\rho_N(r) - \rho_{N-1}(r)] \simeq \rho_{\text{HOMO}}(r) \tag{3.27}$$

$$f_k^- = [q_k(N) - q_k(N-1)] \tag{3.28}$$

$$f_k^- = c_k^2 \tag{3.29}$$

The quantity f_k^- is called the condensed Fukui function.[50] It has a single value for each atom, k, in the molecule, and is not otherwise a function of position. The q_ks are net charges on the atoms. In the last equation c_k^2 is simply the square of the atom coefficient in the HOMO. It is also the frontier orbital density in FMO theory. It is the easiest to calculate, since we only need the wave function for the HOMO, which can often be found, at least roughly, from HMO theory.

Equation (3.28) requires all the filled orbitals, followed by a Mulliken (or other) population analysis, for both the reactant molecule and its cation. It has the advanage that the charge, $q_k(N)$, can also be useful. It will be recalled that only soft–soft interactions between two reactants are controlled by the Fukui function. That is, electron transfer, or covalency, is dominant. For hard–hard interactions, the charges on each atom dictate where reaction will occur.[51]

The function $f^-(r)$ requires a complete calculation, but it can be approximated from the HOMO, which is simpler. It gives more detailed information about the stereochemistry of the reaction path. For example, in a theoretical study of the reactions of HCHO with both electrophiles and nucleophiles, it was found that a base would approach the carbon atom from a direction perpendicular to the plane of the formaldehyde molecule.[52] An acid such as H^+ would attach itself to the oxygen atom, as a result of the net charge being more negative on that atom.

An *ab-initio* calculation of the MOs of the NCS^- ion gives the wave function for the HOMO as[53]

$$\phi = 0.855\psi_S + 0.139\psi_C - 0.444\psi_N \tag{3.30}$$

where ψ is a suitable valence shell atomic orbital. Squaring ϕ, we see that ρ_{HOMO} is much larger on S, than on N. Soft electrophiles will react at sulfur. From the total electron density, however, N is more negative than S, $-0.68e^-$ vs. $-0.25e^-$. Hard electrophiles, whose reactions are controlled by electrostatics, will react at N.

Sometimes even a simple diagram of the frontier orbitals will give us a great deal of chemical information. For example, consider the malonaldehyde anion, a model for the important β-diketone anions

$$\underset{1}{\text{O}}\text{=}\underset{2}{\text{CH}}\text{=}\underset{3}{\text{CH}}\text{=}\underset{4}{\text{CH}}\text{=}\underset{5}{\text{O}}$$

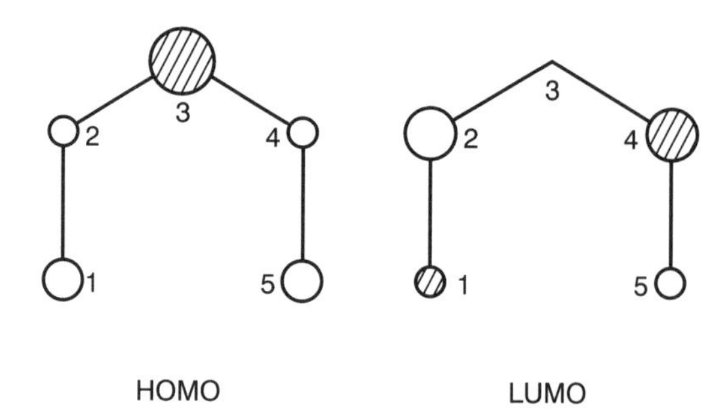

Figure 3.8 The HOMO and the LUMO of malonaldehyde. The shading shows the phases, or signs, and the diameter of the spheres shows the magnitude of the coefficient in the MO. Reprinted with permission from R.C. Haddon, J. Am. Chem. Soc., *102*, 1807 (1980). © 1980 American Chemical Society.

The HOMO and the LUMO are both π-type orbitals for which an HMO calculation has been made.[54] Figure 3.8 shows a diagram commonly used to show such orbitals. The diameters of the spheres represent the coefficients of each atomic orbital in the MO. We readily see that carbon atom 3 is the reactive site for reaction with soft electrophiles. We can assume that the oxygen atoms will be more negative than any of the carbon atoms, and will react with hard electrophiles.

The LUMO diagram, which gives f_k^+, shows that carbon atoms 2 and 4 will be the reactive sites. Even hard bases, such as OH^-, will react at these positions, since these are carbonyl carbons and will be more positive than the central carbon atom. Such qualitative conclusions are supported by detailed calculations on maleimide, a cyclic molecule related to the malonaldehyde ion.[55] In this case f^+ was calculated from

$$f^+(r) = [\rho_{N+1}(r) - \rho_N(r)] \simeq \rho_{\text{LUMO}} \tag{3.31}$$

The function $f^+(r)$, shown as density contour lines, $\rho_{\text{LUMO}}(r)$ and f_k^+ all gave the same predictions about reaction sites.

By simply multiplying f^+, or f^-, by the global softness, σ, one obtains the local softness, $\tilde{\sigma}$. Since softness is an index of molecular reactivity, one might hope to rank-order a series of similar molecules in terms of reactivity to a given reagent. This has been attempted in the case of electrophilic substitution of substituted benzenes, C_6H_5X.[56] It was found that $f^-(r), f_k^-$ and $\rho_{\text{HOMO}}(r)$ all gave correct answers for the relative *ortho,para-meta* reactivity, except for X = CN and NO_2 This is not too surprising since c_k^2, i.e., frontier orbital theory, also fails in these cases. It is necessary to include the next-highest MO in order to get the correct answer.[57]

More importantly, it was found that $\tilde{\sigma}$ did not give the correct order of intermolecular activity. Table 3.16 shows why this is the case. The relative values of η do not correlate with reactivity, but the values of χ do. In frontier orbital theory, it is the ionization potential, or the energy of the HOMO, which is most closely related to overall reactivity in reactions such as aromatic electrophilic substitution.[57] In other cases it is the variations in electron affinity which explain the experimental results. Generally speaking, it is usually better to use both I and A, as required by Equation (3.1).

It is worth noting that another function, the molecular electrostatic potential[58] (MEP) was tested as an index of intermolecular and intramolecular activity and found to give the correct order in all cases.[56] It is the sum of the potentials due to the nuclei and all the electrons at each point in the molecule. It has been widely used as an index of reactivity, but requires considerable effort to calculate.

A condensed form of the MEP could be used, if we knew the net charge on each atom with certainty. But finding these charges, q_k is a difficult, if not impossible, task since is is not well defined. The commonly used Mulliken population analysis has obvious deficiencies. Several dozen alternative methods, both theoretical and experimental, have been proposed.[59] Density functional theory also offers a method, which has some advantages.

One approach would be the calculation of ΔN, which for a diatomic molecule gives $q_k = \pm\Delta N$.[60] But this can only be considered a zero-order approximation; it is necessary to take into account the further changes that occur as the interacting systems approach each other more closely. This has been done and presented as the electronegativity equalization method, or EEM.[61] The key equation is given by

$$\chi_{ks} = (\chi_k^0 + \Delta\chi_k) + 2(\eta_k^0 + \Delta\eta_k)q_k + \sum_l q_l/R_{kl} \tag{3.32}$$

χ_k^0 and η_k^0 are the EN and hardness of the free atom; $\Delta\chi_k$ and $\Delta\eta_k$ are the changes due to changing size and shape of the atom in the molecule; q_l/R_{kl} is the potential of the shielded atom l on the kth atom. We also have

$$\chi_k = \chi_l = \chi_m \ldots \qquad \text{and} \qquad \sum_l q_k = 0 \tag{3.33}$$

If we have M atoms, then we have M simultaneous equations, which can be solved for the various q_ks

It would be difficult to calculate $\Delta\chi$ and $\Delta\eta$ for large systems. Instead they are modeled by calculations on small molecules by assuming that Equation (3.33) is valid and that the q values can be found by some population analysis. The values for $\Delta\chi$ and $\Delta\eta$ for each kind of atom are then assumed to be transferable. The EEM method has the advantage that it can be applied to very large systems. Besides the atom charges, one also finds the global EN, the hardness, and the Fukui function for each atom.[62]

IMPROVING THE ENERGY FROM AN APPROXIMATE WAVE FUNCTION

An approximate wave function for a given system will not give the correct electron density, nor will it give the correct electronic chemical potential, constant everywhere. Instead it will give a potential, μ, which has positive and negative deviations from the average value, $\bar{\mu}$, at various points. An improved density and energy can be had by transferring density from regions where μ is too positive to regions where μ is too negative.

The key equations are familiar

$$\Delta N = \frac{(\mu_D - \mu_C)}{2(\eta_C + \eta_D)} \tag{3.34}$$

$$\Delta E = \frac{-(\mu_D - \mu_C)^2}{4(\eta_C + \eta_D)} \tag{3.35}$$

A simple example is provided by a hydrogen atom in a weak electric field. The chemical potential is given by μ^0, that of an unperturbed H atom, plus Fz, where F is the strength of the field along the z axis. The regions C and D correspond to $-z$ and $+z$. For any two points, $(\mu_D - \mu_C) = (z_D - z_C)F$.

To obtain the energy lowering due to charge transfer in Equation (3.35), we find the averages

$$\Delta E = \frac{-\langle (z_D - z_C)^2 \rangle F^2}{8\eta} \tag{3.36}$$

where the integration is over z from 0 to ∞, or 0 to $-\infty$, for D and C, respectively. We find $\langle (z_D - z_C)^2 \rangle$ to be 3.125 in atomic units. Assuming that $\eta = (I - A)/2$ for the H atom, we have $\eta = 0.236$ a.u. Accordingly

$$\Delta E = -\frac{3.125F^2}{8(0.236)} = -1.655F^2 = -\frac{\alpha F^2}{2} \tag{3.37}$$

so that the polarizability, α, is calculated as 3.31 a.u., compared with the exact value of 4.50 a.u. We can lower the energy further by transferring charge from larger values of z to smaller values in region D, and the reverse for region C. This improves α to 3.97 a.u.

However, there is a problem with this kind of calculation. It does not seem reasonable to use the approximation $\eta = (I - A)/2$ for problems of this kind. We are not adding or removing charge in the sense of adding or removing electrons, but merely rearranging a constant amount of charge. There is no change in electron–electron repulsion, obviously, so only the change in kinetic energy with ρ should be considered. But this functional dependence is not easily found.[63]

The same approach as the above has been applied to atoms and some very simple molecules.[64] The difference in μ is approximated by $(\varepsilon - \bar{\varepsilon})$, where ε is the local orbital energy. It was found that good results were obtained if it was assumed that $\eta = \langle (t + v_e) \rangle$, where t is the local one-electron kinetic energy and v_e is the electron repulsion potential. Such an assumption can only be partially justified.

The cases of slightly perturbed particles in boxes and harmonic oscillators has also been treated.[65] Good results were obtained if the hardness was simply equated to the kinetic energy. It is of some interest that DFT can be applied to particles other than electrons. The similarity of Equation (3.35) to the energy correction of second-order perturbation theory is also worth noting

$$\Delta E = \sum_k \frac{\langle \psi_0 | H' | \psi_k \rangle^2}{(E_0 - E_k)} \simeq \sum_k \frac{\langle \psi_0 | H' | \psi_k \rangle^2}{\bar{E}} \tag{3.38}$$

H' is the perturbation and $\bar{E}$ is the average excitation energy between the ground state, ψ_0, and the various excited states, ψ_k. Only the lowest one or two states are important, in many cases.

For heavy atoms, and for complex systems such as solids, it is common practice to replace the inner shells by an effective potential, the pseudopotential (PP), which has the same influence on the outer parts of the atom as the real potential. The PP is used in a region with a surface boundary beyond which the real potenial is used. Since we know something about the behavior of wave functions and orbital energies at large distances from the atom, the results for the outer region are known to some extent.

Accordingly, it is important to match the orbitals and energies generated by the PP in the pseudo-atom region to the corresponding properties in the true atom region. It has been found that considerable improvement occurs if the hardness is one of the conserved properties.[66] That is, the total energy of the atom and pseudo-atom must be the same up to the second order for small changes in the frontier orbital occupancy. The improvement lies mainly in the transferability of the PP. The same PP can be used for a given atom in a variety of environments.

SOLVATION EFFECTS

So far the concepts derived from DFT have been applicable to isolated chemical systems, i.e., to gas-phase molecules, atoms and ions. Actually some of the applications discussed were based on results in solution, with only minor comments on the effects. Since solvation energies will always be important for chemical reactions, it is time to examine solvation in more detail. There are two possible procedures: one is to use the gas-phase theory first, and then to modify the conclusions by a separate analysis of solvation effects using well-developed solvation theory. The second is to include the solvent as part of the system, and then to develop the concepts from DFT.

We will illustrate the latter method by taking the important case of aqueous solutions.[67] Just as we have I and A in the gas phase, we have the corresponding properties in solution

$$M(aq) = M^+(aq) + e^-(g) \qquad I' \tag{3.39}$$

$$M(aq) + e^-(g) = M^-(aq) \qquad A' \tag{3.40}$$

It is convenient to define I' and A' as the potentials of the standard Gibbs free energy changes in Equations (3.39) and (3.40). The free energy of the electron at rest in the gas is set equal to zero.

The direct measurement of I' and A' is not possible, but they can be calculated in several independent ways. Electrochemical methods depend on knowing the absolute potential of the hydrogen electrode.

$$H^+(aq) + e^-(g) = \tfrac{1}{2}H_2(g) \qquad E^0_H \tag{3.41}$$

There is now general agreement on a value near 4.5 V at 25 °C.[68] This value enables us to find the free energy of hydration of the proton (−260 kcal/mol), and from this the free energies of hydration of many other ions.[67,69] Combined with gas-phase data on I and A, many values of I' and A' can be calculated.[67]

Since a thermochemical cycle is used, these would be adiabatic values. It is of interest that vertical values can be found experimentally by the technique of photoelectron emission spectroscopy.[70] These are 1 or 2 V higher because they do not include the reorganization energy of the solvent around the product.

By combining elecrochemical methods with other data, it has been possible to obtain values of I' and A' for large numbers of both inorganic[71,72] and organic[73] molecules. The determination of half-wave potentials, $E_{1/2}$, by cyclic voltammetry has been particularly useful for organic molecules.[74] Even when "$E_{1/2}$" is irreversible, the variation for a series of related molecules is about the same as for the reversible values.[75] These results for organic molecules are usually obtained in solvents such as acetonitrile or dimethyl sulfoxide. Fortunately there is a large body of information on the free energies, or enthalpies, of transfer of ions from one polar solvent to another.[76]

Table 3.19 gives a small sample of the results which may be obtained. For various practical reasons, it is usually difficult to measure both I' and A' for the same species, so that μ' and η' are not known for most systems. However, this is not as serious a problem as might be supposed. It may be seen that values of I' are 2–4 eV less than the gas-phase values I for the same molecule. Also A' values are 2–4 eV greater than the gas phase A values. The larger the neutral species,

Table 3.19 I' and A' for Some Sample Systems, Water Solution, 25 °C[a]

M	I' [eV]	M	A' [eV]
$C_6H_5NH_2$	4.53	CH_3	3.73
$(CH_3)_3N$	5.21	PH_2	4.18
$(CH_3)_2S$	6.11	C_6H_5	4.27
$(CH_3)_3P$	6.34	O_2	4.31
C_6H_6	6.56	C_6H_5S	5.36
$(C_2H_5)_2O$	6.72	C_6H_5O	5.42
$(CH_3O)_3P$	6.75	NH_2	4.72
CH_3OH	7.25	I	5.64
H_2O	7.94	OH	6.22
CH_3NO_2	8.65	CN	7.02
CH_3CN	9.35	F	7.88

[a] See Reference 67 for data. Also see A. Bagno and G. Scorrano, J. Am. Chem. Soc., *110*, 4577 (1988) for new data on pK_as.

the smaller is the difference. For example, for OH and C_6H_5S radicals, using Tables 3.11 and 3.19, we find $(A' - A)$ equal to 4.39 and 2.79 eV, respectively.

Obviously the differences between I and I' and A and A' are almost entirely the result of strong solvation of the ions, M^+ or M^-. It is easier to form ions in any solvent, than in the gas. Figure 3.9 now shows the effect of this on μ' and η' in an energy diagram. On going from the gas to solution, A becomes greater and I becomes less so that the difference between them, the energy gap or $2\eta'$, becomes less. For large moleculs the change from I to I' will be equal and opposite to the change from A *to* A'. Therefore the numerical value of μ' (or χ') will be the same as that of μ (or χ). This will not be the case, however, for small molecules, particularly the hydrogen atom.[77]

Therefore, for a possible transfer of an electron from C to D when both are large, we can use either χ_C or χ'_C, compared with χ_D or χ'_D, to decide which molecule is the donor and which is the acceptor. I' and A' also give the thermodynamic one-electron oxidation potential and reduction potential of the molecule.

$$E^0_{0\chi} = -I' + 4.50\,\mathrm{V} \qquad E^0_{\mathrm{red}} = A' - 4.50\,\mathrm{V} \tag{3.42}$$

Such potentials are important in deciding on the feasibility of single-electron transfer (SET) between different molecules. SET mechanisms are very important in both organic and inorganic chemistry.[78]

The case for partial transfer of electrons from D to C is quite different. The small value of η' in Figure 3.9 is not a reliable number to use for calculating ΔN. The reason is that I and A can be related to the orbital energies of the HOMO and LUMO in the gas phase, but I' and A' cannot be treated in the same way in solution. That is, solvation does not raise the energy of the HOMO and lower

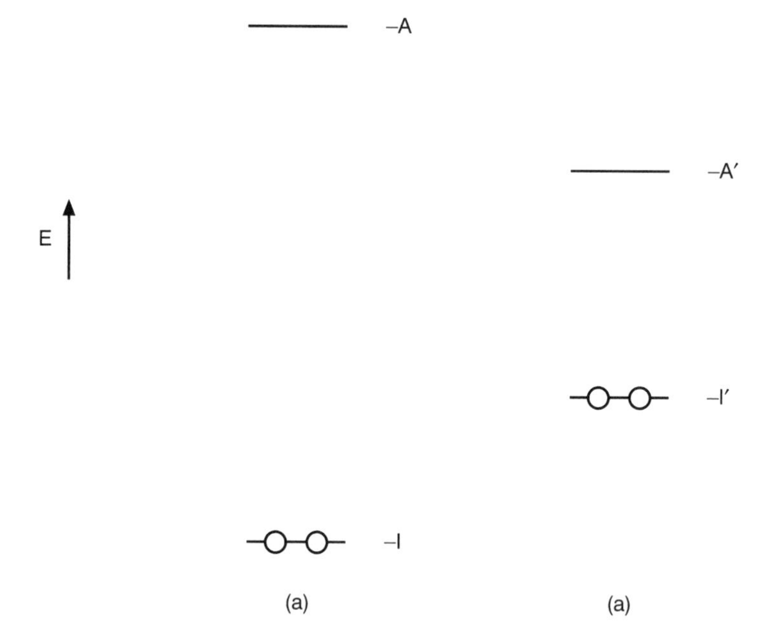

Figure 3.9 (a) Orbital energies of HOMO and LUMO in the gas phase; (b) apparent orbital energies in solution. see text.

that of the LUMO. The orbital energies, in fact, are relatively unchanged. The solvent lowers the energy of all orbitals a small amount for M, and a much larger amount for M^+ or M^-.

The evidence for this can be readily seen by measuring the shifts in the vis–UV spectra of molecules on going from the gas phase to solution. Such shifts are usually quite small. Even for charge transfer bands, where the polarity of the molecule changes, the shifts are less than 1 eV.[79] Therefore η' correctly predicts that a molecule is soft in reactions such as

$$2M(aq) = M^+(aq) + M^-(aq) \tag{3.43}$$

but η is a better measure of ΔN, even in solution, if ΔN is small.

Consiering some of the other uses of the energy gap, it seems clear that if the gas-phase gap is relatively unchanged in solution, then η is the correct measure. For example, the use of the hardness as a measure of stability will be valid for $(I - A)$, but not $(I' - A')$, except for reactions such as (3.43). The polarizability of a molecule also is much the same in solution as it is in the gas phase.

As an interesting side comment, a statistical study has been made of the properties of solvents which determine their solvating abilities.[80] It was found that the orbital energies of the HOMO and the LUMO for the isolated solvent molecules are important determinants. There is an empirical rule that soft solutes

Table 3.20 Experimental ΔH_{het}, ΔH_{homo} and ΔG_{ET} for Organic Free Radicals

R	ΔH_{het} [kcal/mol]	ΔH_{homo} [kcal/mol]	ΔG_{ET}
9-Phenyl-xanthyl	42	16	27
Benzyl	112	62	50
t-Butyl	120	72	48

dissolve in soft solvents, and hard solutes in hard solvents. But there has been little attempt to quantify this in terms of DFT.

As mentioned, there are cases where μ' and η' are the correct measures for reactions in solution. Let R be an organic free radical containing trivalent carbon. Then we have[81]

$$R^+ + R^- \underset{\Delta H_{het}}{\longrightarrow} R_2 \underset{\Delta H_{homo}}{\longleftarrow} 2R \tag{3.44}$$

$$\Delta H_{het} - \Delta H_{homo} \simeq \Delta G_{ET} = 2\eta \tag{3.45}$$

where ΔG_{ET} is the sum of the free energies for the one-electron oxidation of R, and the one-electron reduction of R. The quantity $T\Delta S_{ET}$ is known to be small, and ΔG_{ET} is the difference $(I' - A')$. Some experimental results for different R are given in Table 3.20.

The 9-phenylxanthyl radical is a resonance-stabilized triphenylmethyl analog. The corresponding carbonium ion and carbanion are also stabilized and can be prepared in sulfolane, so that ΔH_{het} can be directly measured.[82] The data for benzyl and t-butyl are obtained by measuring the reduction and oxidation potentials of the radicals in acetonitrile.[83] The results show that $C_6H_5CH_2$ and $(CH_3)_3C$ are much harder than the 9-phenylxanthyl radical (the latter is just one of several studied with similar properties[82]). The solution hardnesses are then responsible for the difficulty in forming the ions in the benzyl and t-butyl cases, and the stability of the ions in the resonance-stabilized cases. The effect of the small hardness in the latter cases also is evident in the small bond energy for homolytic dissociation.

Equations (3.44) and (3.45) are applicable to the symmetrical case shown in reaction (3.43), but they also apply to the general case

$$R^+ + X^- \rightarrow RX \leftarrow R + X \tag{3.46}$$

where X is any radical. For a pair of radicals, such as R and X, the effective hardness is given by[81]

$$\eta = (I_{min} - A_{max})/2 \tag{3.47}$$

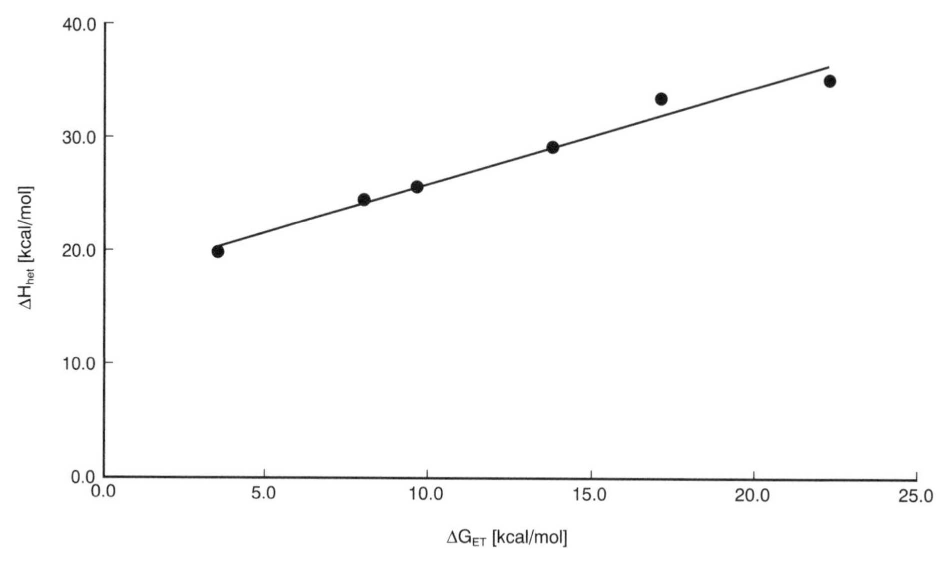

Figure 3.10 Correlation of $\Delta H_{\rm het}$ of carbon–carbon bonds with $\Delta G_{\rm ET}$ for the trityl cation reacting with substituted fluorenide anions. Energies are in kcal/mol. After Reference 81.

In the example given, $I_{\rm R} < I_{\rm X}$ and $A_{\rm X} > A_{\rm R}$, so that the heterolytic bond breaking leads to R^+ and X^-, rather than R^- and X^+.

Equation (3.45) has been tested for the breaking of different carbon–carbon, carbon–oxygen, carbon–nitrogen and carbon–sulfur bonds.[81] For a given R, X was varied. Both R^+ and X^- were cases of stable ions, by resonance or otherwise. The test consisted of plotting experimental values of $\Delta H_{\rm het}$ against $\Delta G_{\rm ET}$ in sulfolane. Figure 3.10 shows the results for carbon–carbon bond breaking. The cation is the trityl ion, $(C_6H_5)_3C^+$, and the anions are substituted fluorenides.

The good linear relationship shown means that $\Delta H_{\rm homo}$ either varies as $\Delta H_{\rm het}$ does, or is constant. The latter turns out to be the case. The main point is that the effective hardness is a good measure of the energy needed to form the ions. This is most useful, since the customary practice in organic chemistry is to use the stabilities of R^+, R^- and R as criteria for ranking the reactivity of their tetracovalent precursors.

For example, good linear relationships are found between $\Delta H_{\rm het}$ and $(pK_a - pK_{R^+})$ for related families of R and X.[84] The pK_a and pK_{R^+} refer to the equilibria

$$HX = H^+ + X^- \qquad K_a \tag{3.48}$$

$$R^+ + H_2O = ROH + H^+ \qquad K_{R^+} \tag{3.49}$$

where reaction (3.48) is measured in dimethyl sulfoxide and (3.49) in strongly acidic water. These equilibria have long been used to estimate the stability of various organic cations and carbanions. Other chemical properties of organic radicals and ions are also related to their redox properties.[84]

REFERENCES

1. For enlightening diagrams of MOs, see W.L. Jorgensen and L. Salem, *The Organic Chemist's Book of Orbitals*, Academic Press, New York, 1973.
2. R.G. Pearson, *Acc. Chem. Res.*, **4**, 152 (1971).
3. For example, see R.G. Pearson and H.B. Gray, *Inorg. Chem.*, **2**, 358 (1963).
4. R.G. Pearson, *Inorg. Chem.*, **23**, 4675 (1984).
5. T. Ziegler, V. Tschinke and C. Ursenbach, *J. Am. Chem. Soc.*, **109**, 4825 (1987); J. Li, G. Schreckenbach and T. Ziegler, *J. Phys. Chem.*, **98**, 4838 (1994).
6. R.G. Pearson in *Bonding Energetics in Organometallic Compounds*, T. Marks, Ed., ACS Symposium Series 428, Washington, DC, American Chemical Society, 1990, pp. 251–262.
7. C.A. Tolman, *J. Am. Chem. Soc.*, **96**, 2780 (1974).
8. K. Kitaura, S. Sakaki and K. Morokuma, *Inorg. Chem.*, **20**, 2292 (1981); T. Ziegler, ibid., **24**, 1547 (1985).
9. R.G. Pearson, *Inorg. Chim. Acta*, **198–200**, 781 (1992).
10. T. Ziegler, *Inorg. Chem.*, **25**, 2721 (1986).
11. D. Marynick, *J. Am. Chem. Soc.*, **106**, 4064 (1984).
12. M. Guerra, D. Jones, G. Distefano, A. Foffani and A. Modelli, *J. Am. Chem. Soc.*, **110**, 375 (1988).
13. W. Kutzelnigg, *Angew. Chem., Int. Ed. Engl.*, **23**, 272 (1984).
14. R.G. Pearson, *J. Chem. Ed.*, **64**, 561 (1987).
15. J.P. Collman and L.S. Hegedus, *Principles and Applications of Organotransition Metal Chemistry*, University Science Books, Mill Valley, CA, 1980, Chapter 4.
16. R.G. Pearson, *Inorg. Chem.*, **27**, 734 (1988).
17. W. Walper and H. Keim, *Z. Phys. Chem.* (Munich), **113**, 207 (1978).
18. S. Sakaki and M. Ieki, *J. Am. Chem. Soc.*, **113**, 5068 (1991).
19. C. DiBugno, M. Pasquale, P. Leoni, P. Sabatino and D. Braga, *Inorg. Chem.*, **28**, 1390 (1989).
20. For theoretical calculations see M.R.A. Blomberg and P.E.M. Siegbahn, *J. Chem. Phys.*, **78**, 986, 5682 (1983); J.J. Low and W.A. Goddard, *Organometallics*, **5**, 609 (1986); T. Ziegler, V. Tschinke and A. Becke, *J. Am. Chem. Soc.*, **109**, 1351 (1987).
21. R.G. Pearson and P.E. Figdore, *J. Am. Chem. Soc.*, **102**, 1541 (1980); R.G. Pearson in *Nucleophilicity*, J.M. Harris and S.P. McManus, Eds., ACS Symposium Series 215, American Chemical Society Washington, DC, 1987, pp. 233–246.
22. M.R.A. Blomberg, J. Schule and P.E.M. Siegbahn, *J. Am. Chem. Soc.*, **111**, 6156 (1989).
23. C.W. Bauschlicher, Jr., *Chem. Phys. Lett.*, **142**, 71 (1988).
24. R.D. Gillard, J.A. Osborn and G. Wilkinson, *J. Chem. Soc.*, 4107 (1965).
25. J.P. Collman and M.R. Laury, *J. Am. Chem. Soc.*, **96**, 3019 (1974).
26. M. Kubota, *Inorg. Chim. Acta*, **7**, 195 (1973).
27. R.G. Pearson, *J. Org. Chem.*, **54**, 1423 (1989).
28. A. Tachibana and K. Namamura, *J. Am. Chem. Soc.*, **117**, 3605 (1995).
29. J. Sauer, H. Wiest and A. Mielert, *Chem. Ber.*, **97**, 3183 (1964).
30. J.M. Tedder, *Angew. Chem., Int. Ed. Ebgl.*, **21**, 401 (1982); B. Giese, ibid., **22**, 753 (1983).

31. K.H. Becker, R. Kurtenbach and P. Wiesen, *J. Phys. Chem.*, **99**, 5986 (1995).
32. S.G. Lias, J.E. Bartmess, J.F. Liebman, J.L. Holmes, R.D. Levin and W.G. Mallard, *J. Phys. Chem. Ref. Data*, **17**, Suppl. No. 1 (1988).
33. R.G. Pearson, *J. Mol. Struct. (Theochem.)*, **255**, 261 (1992).
34. K. Heberger and A. Lopata, *J. Chem. Soc., Perkin Trans.*, **2**, 91 (1995).
35. R.W. Quandt and J.F. Hershberger, *Chem. Phys. Lett.*, **206**, 355 (1993).
36. R.J. Balla, H.H. Nelson, and J.R. McDonald, *Chem. Phys.*, **109**, 101 (1986).
37. J.J. Russell, J.A. Seetula, D. Gutman, F. Danis, F. Caralp, P.D. Lightfoot, R. Lesclaux, C.F. Melius and S.M. Senkan, *J. Phys. Chem.*, **94**, 3277 (1990).
38. A. Masaki, S. Tsunashima and N. Washida, *J. Phys. Chem.*, **99**, 13126 (1995).
39. R. Atkinson, *Chem. Rev.*, **86**, 69 (1986); J.P.D. Abbott and J.G. Anderson, *J. Phys. Chem.*, **95**, 2382 (1991).
40. G. Klopman, *J. Am. Chem. Soc.*, **90**, 223 (1968).
41. K. Munger and H. Fischer, *Int. J. Chem. Kinet.*, **17**, 809 (1985).
42. L.C. Allen, E.T. Egolf and C. Liang, *J. Phys. Chem.*, **94**, 5602 (1990).
43. A. Streitwieser, Jr., *Molecular Orbital Theory for Organic Chemists*, John Wiley, New York, 1961, Chapter 11.
44. G.W. Wheland, *J. Am. Chem. Soc.*, **64**, 900 (1942).
45. N.S. Isaacs, *Physical Organic Chemistry*, Longman, Harlow, UK, 1987, Chapter 10.
46. S. Clementi, F. Genel and G. Marino, *Chem. Commun.*, 498 (1967).
47. K. Fukui, T. Yonezawa and H. Shingu, *J. Chem. Phys.*, **20**, 722 (1952).
48. Z. Zhou and R.G. Parr, *J. Am. Chem. Soc.*, **112**, 5720 (1990).
49. R.F.W. Bader, *Can. J. Chem.*, **40**, 1164 (1962).
50. W. Yang and W.J. Mortier, *J. Am. Chem. Soc.*, **108**, 5708 (1986).
51. For a review of orbital control and charge control, see G. Klopman, *Chemical Reactivity and Reaction Paths*, John Wiley, New York, 1974, Chapter 4.
52. C. Lee, W. Yang and R.G. Parr, *J. Mol. Struct. (Theochem)*, **163**, 305 (1988).
53. K.A. Jorgensen and S. Lawesson, *J. Am. Chem. Soc.*, **106**, 4687 (1984).
54. R.C. Haddon, *J. Am. Chem. Soc.*, **102**, 1807 (1980).
55. F. Méndez, M. Galván, A. Garritz, A. Vela and J. Gásquez, *J. Mol. Struct. (Theochem.)*, **277**, 81 (1992).
56. W. Langenecker, K. Demel and P. Geerlings, *J. Mol. Struct. (Theochem.)*, **234**, 329 (1991); idem, ibid., **259**, 317 (1992).
57. J. Fleming, *Frontier Orbitals and Organic Chemical Reactivity*, John Wiley, New York, 1976, Chapter 3.
58. R. Bonaccorsi, E. Scrocco and J. Tomasi, *J. Chem. Phys.*, **52**, 5270 (1970).
59. For a review, see J. Meister and W.H.E. Schwarz, *J. Phys. Chem.*, **98**, 8245 (1994). See also J. Cioslowski and S.T. Mixon, *J. Am. Chem. Soc.*, **115**, 1084 (1993).
60. D. Bergman and J. Hinze, *Structure and Bonding*, **66**, 145 (1987).
61. W.J. Mortier, S.K. Ghosh and S.J. Shankar, *J. Am. Chem. Soc.*, **108**, 4315 (1986); K.A. Van Genechten, W.J. Mortier and P. Geerlings, *J. Chem. Phys.*, **86**, 5063 (1987).
62. B.G. Baekelandt, W.J. Mortier, J.L. Lievens and R.A. Schoonheydt, *J. Am. Chem. Soc.*, **113**, 6730 (1991).
63. R.G. Parr and W. Yang, *Density Functional Theory of Atoms and Molecules*, Oxford University Press, New York, 1989, Chapter 6.
64. R.G. Pearson and W.E. Palke, *Int. J. Quantum Chem.*, **38**, 103 (1990).
65. R.G. Pearson, *Theor. Chim. Acta*, **78**, 281 (1991).
66. M. Teter, *Phys. Rev. B*, **48**, 5031 (1993).
67. R.G. Pearson, *J. Am. Chem. Soc.*, **108**, 6109 (1986).

68. S. Trasatti, *Pure Appl. Chem.*, **58**, 955 (1986).
69. K.W. Frese, Jr., *J. Phys. Chem.*, **93**, 5911 (1989).
70. I. Watanabe, *Anal. Sci.*, **10**, 229 (1994).
71. D.M. Stanbury, *Adv. Inorg. Chem.*, **33**, 69 (1989).
72. D.E. Richardson, *Inorg. Chem.*, **29**, 3213 (1990).
73. For example see F.G. Bordwell and X.-M. Zhang, *Acc. Chem. Res.*, **26**, 510 (1993); T. Heinis, S. Chowdhury, S.L. Scott and P. Kerbarle, *J. Am. Chem. Soc.*, **110**, 400 (1988); R.S. Ruoff, K.M. Kadish, P. Boulas and E.M. Chen, *J. Phys. Chem.*, **99**, 8843 (1995).
74. R. Breslow, *Pure Appl. Chem.*, **40**, 493 (1974).
75. L.I. Miller, G.D. Nordblum and E.A. Mayeda, *J. Org. Chem.*, **37**, 916 (1972).
76. Y. Marcus, M.J. Kamlet and R.W. Taft, *J. Phys. Chem.*, **92**, 3613 (1988).
77. R.G. Pearson, *Chem. Rev.*, **85**, 41 (1985).
78. L. Eberson, *Adv. Phys. Org. Chem.*, **18**, 79 (1982); W. Kaim, *Acc. Chem. Res.*, **18**, 160 (1985).
79. C. Reichardt, *Angew. Chem., Int. Ed. Engl.*, **4**, 29 (1965).
80. M. Chastrette, M. Rajzmann, M. Chanon and K.F. Purcell, *J. Am. Chem. Soc.*, **107**, 1 (1985).
81. P.K. Chattaraj, A. Cedillo, E.M. Arnett and R.G. Parr, *J. Org. Chem.*, **60**, 4707 (1995).
82. E.M. Arnett, R.A. Flowers, II, A.E. Meekhof and L. Miller, *J. Am. Chem. Soc.*, **115**, 12603 (1993).
83. D.D.M. Wayner, D.J. McPhee and D. Griller, *J. Am. Chem. Soc.*, **110**, 132 (1988).
84. E.M. Arnett and R.T. Ludwig, *J. Am. Chem. Soc.*, **117**, 6627 (1995).

4 The Principle of Maximum Hardness

INTRODUCTION

From what has been said already, it is obvious that the hardness of a chemical system plays a major role in determining its stability or reactivity. Alternatively, we can use the HOMO–LUMO energy gap as a criterion. If stability is desired, then it is advantageous to have a large energy gap, or a high value of the hardness. If reactivity is desired, then a small gap or hardness is desirable.

Examples have already been given in Chapter 2 showing that soft molecules are more reactive than similar, but harder, molecules. There is much more evidence consistent with the statement that increasing hardness goes with increasing stability. For example, theoretical calculations almost always indicate that the most stable structure has the largest HOMO–LUMO gap. A simple case is shown in Figure 4.1, where the MO energy diagram is given for CH_4 in both its stable tetrahedral structure and an unstable planar form.

The smaller gap in the latter case arises because the p_z orbital of carbon is removed from bonding while still occupied. It rises in energy and becomes the HOMO. At the same time a linear combination of hydrogen 1s orbitals is removed from an anti-bonding MO of tetrahedral methane and becomes non-bonding. It goes down in energy and becomes the LUMO. In fact, since we know that the tetrahedral structure of CH_4 has the maximum overlap of the valence orbitals of C and H, we can assume that any change in structure will result in energy effects similar to Figure 4.1. The hardness will get less and the stability will decrease.

The same results are found in all of the cases where structures are predicted by simple rules, such as the valence-shell electron-pair repulsion theory.[2] There are also many examples involving more complicated molecules. Table 4.1 shows the results of some calculations by the extended Hückel method for the *closo*-borane anions.[3] For $B_6H_6^{2-}$ and $B_7H_7^{2-}$ the HOMO–LUMO gap is given for both the stable structure and unstable ones. The stable structures, octahedral for $B_6H_6^{2-}$ and pentagonal bipyramidal for $B_7H_7^{2-}$, have by far the largest gaps. For $B_8H_8^{2-}$ the gap has been calculated for three structures, for any of which no strong preference is seen. In agreement with this, the ion has been found to be fluxional.

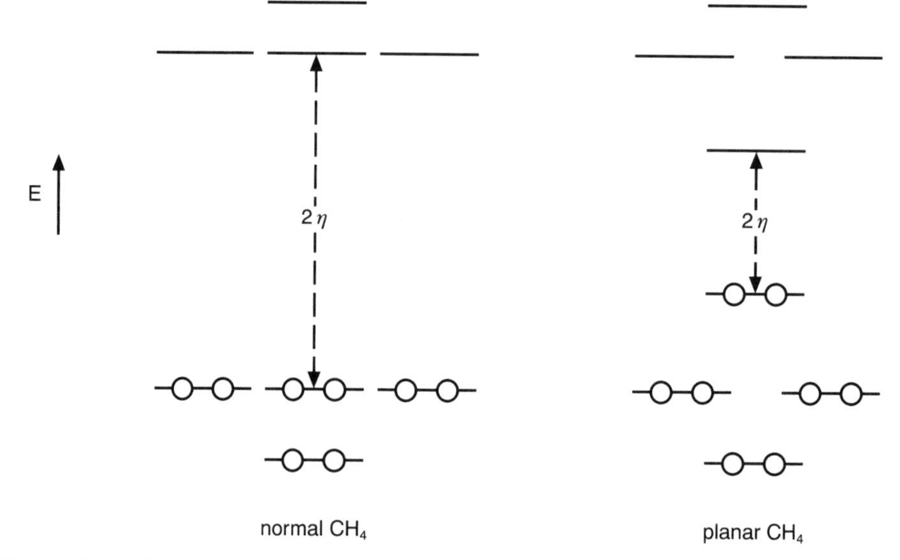

Figure 4.1 MO energy diagram for methane in its stable tetrahedral form and in the unstable planar form.

An examination of orbital interactions when two partial systems approach each other is very informative.[4] Assume that the HOMO and the LUMO play the dominant role in these interactions. Figure 4.2(a) shows the interaction of the frontier orbitals. The lower-energy orbital, the HOMO, goes down in energy, and the LUMO goes up. There is a net energy lowering, and the gap increases. Figure 4.2(b) shows the case where the HOMO is interacting mainly with another filled orbital. Now the HOMO goes up in energy. The LUMO will mix primarily with other empty orbitals, which will lower its energy, as shown. The net effect is that the energy is raised and the HOMO–LUMO gap is diminished.

This is the result for orbital interactions, or covalent bonding. What about ionic bonding? Consider an anion and a cation approaching each other, with a

Table 4.1 Results of Molecular Orbital Calculations for the *closo*-Boranes[(a)]

Borane anion	Idealized geometry	ΔE_{gap} [eV]	Borane anion	Idealized geometry	ΔE_{gap} [eV]
$B_6H_6^{2-}$	O_h	6.6	$B_7H_7^{2-}$	C_{3v}	0
$B_6H_6^{2-}$	D_{3h}	1.1	$B_8H_8^{2-}$	D_{2d}	2.5
$B_7H_7^{2-}$	D_{5h}	5.9	$B_8H_8^{2-}$	D_{4d}	3.2
$B_7H_7^{2-}$	C_{2v}	1.2	$B_8H_8^{2-}$	C_{2v}	2.2

[(a)] From Reference 3.

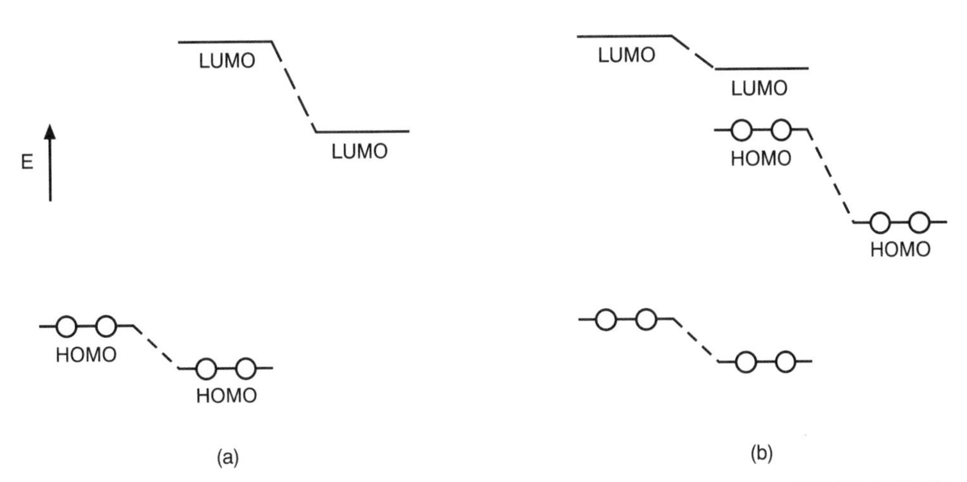

Figure 4.2 Energy changes during the approach of two molecules: (a) HOMO–LUMO interaction; (b) HOMO–second filled orbital interaction.

net decrease in energy. The HOMO will be an atomic orbital on the anion, and the LUMO will be an orbital of the cation, in the usual case. As the ions approach, the potential of the cation will lower the orbital energy of the HOMO, and the potential of the anion will raise the orbital energy of the LUMO. The HOMO and LUMO will move apart, just as in Figure 4.2(a).

These conclusions are drawn for the case where a bond is being formed between two subsystems, but they also apply more generally, as is easily seen in the case of ionic bonding. The HOMO will be a linear combination of anion orbitals, and the LUMO a linear combination of cation orbitals. The largest HOMO–LUMO gap will be found if the arrangement of the ions gives the largest potential at each of the ions, which, in turn, means the minimum energy for the system. Covalent bonding gives the same result, except that orbital overlap is the dominant feature.

Thus covalent and ionic bonding give similar results: the energy decreases when the HOMO–LUMO gap increases. But this conclusion is based on orbital energies, or electronic energy, only. The nuclear repulsion must also be included to get the total energy. This is a major reason why the interaction shown in Figure 4.2 does not continue beyond a certain point.

There is another test that can be applied. Since experimental values of η are available for many systems, we can check to see if η does increase on going from atoms or radicals to stable molecules. Table 4.2 shows some sample results. As expected, the hardness increases on forming the stable product from the unstable reactants. These are not isolated cases; an examination of a large number of reactions shows similar behavior.[5]

Unfortunately there does not seem to be a simple relation between the magnitude of the change in η and the change in energy for the reaction. Partly this results from the neglect of the nuclear repulsion energy, as mentioned. But the

Table 4.2 Changes in η for Simple Reactions[a]

Reaction	η_R [b]	η_P [c]	$-\mu_R$ [b]	$-\mu_P$ [c]
$Na + Cl \rightarrow NaCl$	0.8 eV	4.8 eV	4.4 eV	4.8 eV
$Li + F \rightarrow LiF$	1.0	5.4	4.4	5.9
$Li + H \rightarrow LiH$	2.3	3.8	3.1	4.1
$H + Cl \rightarrow HCl$	4.7	8.0	8.3	4.7
$C + O \rightarrow CO$	4.9	7.9	6.4	6.1
$H + OH \rightarrow H_2O$	5.7	9.5	7.5	3.1
$CH_3 + F \rightarrow CH_3F$	3.2	9.4	6.6	3.2
$CH_3 + Cl \rightarrow CH_3Cl$	3.1	7.5	6.7	3.8
$CH_3 + Br \rightarrow CH_3Br$	3.2	5.8	6.6	4.8
$CH_3 + I \rightarrow CH_3I$	3.4	4.7	6.4	4.9
$Ni + CO \rightarrow NiCO$	3.3	3.6	4.4	4.4
$Cr + 6CO \rightarrow Cr(CO)_6$	3.1	4.5	3.7	3.9
$Cr + 2C_6H_6 \rightarrow Cr(C_6H_6)_2$	3.1	3.3	3.7	2.6
$2H \rightarrow H_2$	6.4	8.7	7.2	6.7
$2N \rightarrow N_2$	7.3	8.9	7.3	7.0
$2F \rightarrow F_2$	7.0	6.3(7.3)[d]	10.4	9.6
$2Na \rightarrow Na_2$	2.3	2.2(?)[d]	2.9	2.7
$2I \rightarrow I_2$	3.7	3.4(3.8)[d]	6.8	6.4

[a] From Reference 5.
[b] Value for reactants
[c] Value for products.
[d] Vertical values for *I* and *A*.

numbers in Table 4.2 show that $\Delta\eta$ is greater for ionic bonding than for covalent bonding. In the latter case it is sometimes necessary to make the correction to the vertical values of *I* and *A*, to obtain an increase in η as predicted.

The increase in the HOMO–LUMO gap due to pure ionic bonding is easily calculated.[5] For NaCl, at the equilibrium distance of 2.36 Å for the diatomic molecule, the electrostatic contribution to η is 6.1 eV. The observed value of 4.8 eV reflects some of the same factors that decrease the gap for covalent bonding. It is interesting to note that ionic bonding should not change the electronic chemical potential, μ. The HOMO is lowered by 6.1 eV, and the LUMO is raised by 6.1 eV, in the case of NaCl. Therefore the midpoint is unchanged.

For covalent bonding, the LUMO is raised more than the HOMO is lowered, as shown in Figure 4.2(a). Hence there is a small positive increase in μ. The results for μ in Table 4.2 are in general agreement with this, but there is no correlation between $\Delta\mu$ and the ΔE of the reaction.

A question may arise as to the experimental values of μ and η for mixtures, such as the reactants in Table 4.1. Consider the case of NaCl again, where the reactants are an equimolar mixture of Na and Cl atoms, which are momentarily prevented from reacting in some way. Removal of an electron from the metastable mixture would certainly cause the formation of Na^+, rather than Cl^+.

Addition of an electron would lead to Cl^-, rather than Na^-. Thus $(I - A)/2$ would be $(5.14 - 3.62)/2$, or 0.7 eV. In general

$$\eta = (I_{\min} - A_{\max})/2; \qquad \mu = -(I_{\min} + A_{\max})/2 \tag{4.1}$$

This intuitive result has been confirmed by a detailed statistical analysis.[6]

Evidence for the correspondence between hardness and stability comes from a consideration of shell structures.[7] Recall that hardness is a consequence of changes in electron–electron repulsion and kinetic energy, as the density function ρ changes. In an atom there is a large change in the kinetic energy when a shell is filled and another electron is added. This causes a large change in η.[4] When a sub-shell is filled, there is a smaller-than-usual change in the interelectronic repulsion, also changing η. The results can be seen in Figure 4.3 which shows the atomic hardness plotted against the atomic number.

The hardness data are from calculations using spin-polarized density functional theory.[8] These results are only in modest agreement with experimental values, but trends are reproduced correctly and, most importantly, the noble gases have valus which are comparable with those of the other elements. This makes it easier to see

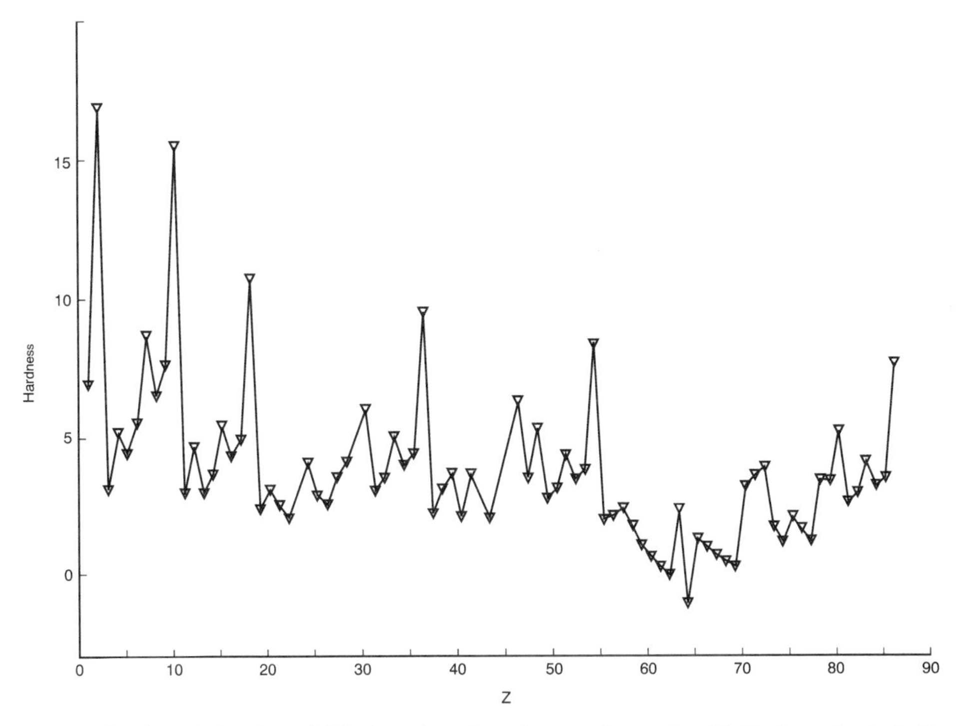

Figure 4.3 Atomic hardness [eV] plotted against the atomic number Z. Shells and sub-shells are determined by the local maxima in hardness. Reproduced with permission from Reference 7.

that η for the atoms is a measure of stability, with the noble gases showing maximum values of η and maximum stability in the sense of chemical inertness.

The special stability of filled sub-shells, or half-filled sub-shells, is also clearly seen. Compare Be with Li and B, or N with C or O. The sub-shell effects are obscured by other factors at large atomic numbers, but are still visible if experimental values are used.

Molecules are also examples of the special stability of filled shells. The simple molecular orbital rule is that a molecule is thermodynamically stable if all the bonding MOs are filled, and all the anti-bonding ones are empty.[9] Other examples are the Octet Rule of Lewis and Langmuir and the 18-Electron Rule of Sidgwick. These relate hardness to stability just as Figure 4.1 does. So do the famous Hückel rule ($4n + 2$), and the Wade–Mingos rule ($2n + 2$).[10]

Parr and Zhou have pointed out that the relationship between hardness and stability can be extended to atomic nuclei.[7] The hardness is defined as half the difference between the highest filled nuclear energy level and the lowest empty one. This difference is greatest for closed-shell nuclei compared with the neighboring open-shell nuclei.[11] Shells can be identified by the "magic numbers" 2, 8, 20, 28, 50, 82 and 126. Hardness seems to be a general measure of stability. Local maxima can be used to locate shell and sub-shell structure.[7]

In quantum mechanical calculations on atoms and molcules, we usually have only approximate electron density functions available. What is the relationship of hardness, η, to ρ? Will it increase to a maximum value as a set of trial basis functions becomes better and better, approaching the true electron density?[4]

Some conclusions can be drawn, if we restrict ourselves to LCAO–MO theory. The eigenfunctions, ϕ_m, are given by

$$\phi_m = \sum C_{ml}\psi_1 \tag{4.2}$$

where the ψs are atomic orbitals. The orbital energies are found from the determinant

$$|H_{ij} - S_{ij}\varepsilon| = 0 \tag{4.3}$$

The conditions for the solutions are that

$$(\partial\varepsilon_m/\partial C_{ml}) = 0; \qquad (\partial E/\partial C_{ml}) = 0 \tag{4.4}$$

All the roots are maxima or minima on the energy-coefficient hypersurface. The lowest root is the absolute minimum, the highest root is the absolute maximum, and the remaining roots are local maxima and minima. The occupied orbitals, in virtually all cases, will have negative curvatures $(\partial^2\varepsilon/\partial C_{ml}^2)$ corresponding to bonding MOs, or zero curvature for non-bonding orbitals. The empty orbitals will have positive curvatures if anti-bonding, or zero curvature. This applies to the HOMO and LUMO, in particular.

The best values of the C_{ml}s define the best wave function, and the best value of ρ, that can be obtained from the selected basis set of AOs. Any change from the best values will cause the HOMO to rise in energy, or be unchanged, and the LUMO will fall in energy, or be unchanged. Thus the energy gap between them is a maximum for the best values of the coefficients, or the "best" electron density function. Usually, of course, this will not be the true density function.

The conclusion that the hardness has a maximum value for the lowest-energy solution can be readily verified in simple cases, such as the Hückel MO theory. Inclusion or omission of overlap integrals has no effect. Examples where all atoms are the same and cases where the atoms and the orbitals are different give the same result. In fact, in simple Hückel theory it can be proved that the most negative value of the π-electron energy is always accompanied by the largest value of the hardness, since both are proportional to the exchange integral, β.[12] However, the proof requires that μ be kept constant.

Provided the self-consistent field (SCF) condition is met, calculations at the Hartree–Fock level also obey the mathematics of Equations (4.3) and (4.4). Therefore the HOMO–LUMO gap should also be a maximum in these cases. Because the solutions are normalized and orthogonal, and because the atomic orbitals are conserved, the coefficients for different MOs are not independent. Therefore wrong coefficients in one orbital will usually lead to wrong coefficients in all orbitals.

We can also change the trial wave function for a system by enlarging the basis set of atomic orbitals. If this is done, what usually happens is that the HOMO is changed only slightly, to a more negative energy, but the LUMO is decreased in energy much more, so that the HOMO–LUMO gap is smaller, even though the total energy is decreased. At the same time, the value of μ becomes more negative, because of the decrease in $\varepsilon_{\mathrm{LUMO}}$. This turns out to be important, as will be seen in the next section.

THE MAXIMUM HARDNESS PRINCIPLE

From much circumstantial evidence of the kind given in the preceding section, it was concluded that "There seems to be a rule of nature that molecules arrange themselves to be as hard as possible".[1] But none of the evidence is rigorous enough or general enough to be conclusive. Fortunately Parr and Chattaraj have given a rigorous and widely general proof of the "Principle of Maximum Hardness", or PMH.[13]

Their proof is based on a combination of statistical mechanics and the fluctuation–dissipation theorem. Since it is not easy to follow, some preliminary discussion is needed. The statistical mechanics part depends on the properties of a grand canonical ensemble. Such an ensemble consists of a large number of systems, each identical with the system of interest. These can exchange energy *and* particles with each other. Therefore the various members of the ensemble can have different numbers of particles and different values of the total energy. These quantities will fluctuate.

The Hohenberg and Kohn theorem applies to ground states at the absolute zero of temperature. Fortunately there is a finite-temperature version of DFT, first proved by Mermin.[14] The equilibrium properties of a grand canonical ensemble are determined by the grand potential, Ω, which is defined as follows:

$$\Omega = E - \mu N - TS \tag{4.5}$$

where S is the entropy. Equilibrium corresponds to a minimum value of Ω.

Equation (4.5) applies either to the case of ordinary thermodynamics, where E, N and μ have their usual meaning, or to the case where E is the electronic energy, N the number of electrons and μ the electronic chemical potential. In either case Mermin showed that the grand potential is a unique functional of the density for a system at finite temperature. Also, the correct density for the system will give a minimum value of Ω. Thus we have a DFT for finite temperature by taking a grand canonical ensemble of the system of interest and calculating its properties.

For example, for ordinary thermodynamics it is found that

$$(\partial N/\partial \mu)_{V,T} = \beta \langle (N - \langle N \rangle)^2 \rangle \tag{4.6}$$

where $\beta = 1/kT$, $\langle N \rangle$ is the average of N over the emsemble, which is constant, and N is the value for each member of the ensemble, which can vary.

For an ensemble of molecular systems we would have[15]

$$(\partial \langle N \rangle/\partial \mu)_{v,T} = \beta \langle (N - \langle N \rangle)^2 \rangle / \langle N \rangle = 2\langle \sigma \rangle \tag{4.7}$$

N is the number of electrons in each molecule, which can vary by transfer between molecules. However, in the cases of interest the average value, $\langle N \rangle$, will be constant, equal to N_0. The restriction of constant v. the nuclear potential, is equivalent to the constant volume, V, of Equation (4.6).

The equilibrium softness of the ensemble is $\langle \sigma \rangle$. Note that this is not the same as the molecular softness, σ_0, but it is related. In either case, the softness depends on the fluctuations of N. Soft systems have large fluctuations. A large value of σ_0 means I is small and A is large. Both gain and loss of electrons are easier in such molecules.

In statistical mechanics we do not measure observables directly. Instead we observe an average over all possible values. The averaging is done by means of a probability distribution function, which in classical mechanics is averaged over all of phase space. Let us compare an equilibrium ensemble with grand potential Ω, and an arbitrary nearby ensemble prepared by a small perturbation, $\Delta\Omega$. Let the equilibrium probability distribution function be f, and that for the nearby ensemble be F. The latter is the equilibrium distribution function for the grand potential $\Omega + \Delta\Omega$. However, at time $t = 0$ the perturbation is turned off in some way. Then F will relax to f, usually in a first-order process characterized by a relaxation time, τ. From the relation between the grand potential and the probability distribution function, we know that

$$F = \frac{\exp(-\beta\Delta\Omega)}{\langle\exp(-\beta\Delta\Omega\rangle} f \tag{4.8}$$

where the averaging is over the members of the ensemble.

Now let A (not the electron affinity) be a dynamic variable whose observed average value will change as F changes to f. This means that A changes with Ω. For small changes we can assume $C\mathrm{A} = \exp(-\beta\Delta\Omega)$, where C is a coupling constant. Then Equation (4.8) becomes

$$F = \langle\mathrm{A}\rangle^{-1}\mathrm{A}f \tag{4.9}$$

In the equilibrium state, the average value is independent of time. Therefore the average value of the observable, $\langle\mathrm{A}\rangle$, is a constant, as is f. Both F and A relax to the final values of f and $\langle\mathrm{A}\rangle$. At $t = 0$ the value of A can be written as $\bar{\mathrm{A}}(0)$, where the overbar indicates an average over the now non-equilibrium ensemble.

All averaging must be done with the μ, v and T of the equilibrium ensemble. Therefore we are comparing equilibrium and non-equilibrium systems of constant μ, v and T. Looking at equations (4.5) and (4.8), we can see that, if μ is constant, the energy of the non-equilibrium ensembles will be higher than that of the equilibrium one. At least this will be the case when changes in entropy are small. Otherwise, it is the free energy which is higher.

As time goes on, the non-equilibrium ensembles will relax towards equilibrium. The excess energy will be degraded to heat, by a mechanism of molecular collision. This means that the fluctuation–dissipation theorem will apply.[15] This theorem dates back to early work by Nyquist[16] and Onsager[17] and has been generalized by Callen and Welton.[18] In essence, it says that small deviations from equilibrium have the same relaxation times, whether they are spontaneous or induced. This follows because the mechanism is the same.

The theorem applies to any observable, A which relaxes from the non-equilbrium value $\bar{\mathrm{A}}$, averaged over the ensemble, to the equilibrium value $\langle\mathrm{A}\rangle$. The relaxation of a spontaneous fluctuation in an equilibrium system is given by the time correlation function, $C(t)$

$$C(t) = \langle(\mathrm{A}(0) - \langle\mathrm{A}\rangle)(\mathrm{A}(t) - \langle\mathrm{A}\rangle)\rangle = \langle\mathrm{A}(0)\mathrm{A}(t)\rangle - \langle\mathrm{A}\rangle^2 \tag{4.10}$$

The correlation function decreases from a maximum value at $t = 0$ to zero at times long compared to the relaxation time, τ.

The fluctuation–dissipation theorem can then be written as

$$\frac{C(t)}{C(0)} = \mathrm{e}^{-t/\tau} = \frac{(\bar{\mathrm{A}}(t) - \langle \mathrm{A} \rangle)}{(\bar{\mathrm{A}}(0) - \langle \mathrm{A} \rangle)} \tag{4.11}$$

This establishes that $(\bar{\mathrm{A}}(t) - \langle \mathrm{A} \rangle)$ is directly proportional to $C(t)$, but it does not give the sign of the constant of propotionality. Some observables in a non-equilibrium system are less than the equilibrium value, and some are greater. For example, consider concentrations of reactants and products in a chemical equilibrium.

To find the constant of proportionality we follow Parr and Chattaraj,[13] who followed Chandler.[15a] We will show that

$$C(t) = \langle \mathrm{A} \rangle (\bar{\mathrm{A}}(t) - \langle \mathrm{A} \rangle) \tag{4.12}$$

First write $\bar{\mathrm{A}}(t)$ in terms of the distribution function F which exists at $t = 0$.

$$\bar{\mathrm{A}}(t) = \int \mathrm{d}r^N \mathrm{d}p^N F(0; r^N p^N) \mathrm{A}(t; r^N p^N) \tag{4.13}$$

The integration is to be over all of phase space. Next replace $F(0; r^N p^N)$ by its equal from (Equation 4.9) and carry out the integration.

$$\bar{\mathrm{A}}(t) = \langle \mathrm{A} \rangle^{-1} \mathrm{A}(0) A(t) \tag{4.14}$$

This result inserted into Equation (4.12) will give $C(t)$ equal to its value in equation (4.10).

Accordingly, it follows that

$$\begin{aligned} \langle \mathrm{A} \rangle (\mathrm{A}(t) - \langle \mathrm{A} \rangle) &= C(t) \\ \langle \mathrm{A} \rangle (\mathrm{A}(0) - \langle \mathrm{A} \rangle) &= C(0) \end{aligned} \tag{4.15}$$

The unknown constant of proportionality in Equation (4.11) is therefore just $\langle \mathrm{A} \rangle$. Now take the observable A to be the softness, σ, which is always positive. Then it must be true that

$$(\bar{\mathrm{A}}(0) - \langle \mathrm{A} \rangle) = (\bar{\sigma} - \langle \sigma \rangle) \geq 0 \tag{4.16}$$

Thus the equilibrium ensemble has the minimum softness, or the maximum hardness. This result is valid for all nearly non-equilibrium distributions which obey Equations (4.8) and (4.7). Presumably it includes all cases of interest, since these are common assumptions in linear response theory.

The restriction to constant μ and v is severe, and greatly limits the usefulness of the proof. However, it is likely that these restrictions can be relaxed somewhat.

The most important feature seems to be that the non-equilibrium system has a higher energy than the equilibrium one, so that the fluctuation–dissipation theorem can be applied. The result of interest in Equations (4.14) and (4.15) is that the constant of proportionality is a quantity that is always positive. Its exact value of $\langle A \rangle$ is not actually used.

An interesting feature of the Parr–Chattaraj proof of the Principle of Maximum Hardness, is that the specific example of chemical softness is not introduced until the last step. The proof should then be valid for many other observables, provided that certain restrictions are met. One requirement is that the observable always has a positive value (or in some cases always a negative one).

Equation (4.7) suggests that the fluctuations in an equilibrium system will offer many examples of the minimum (maximum) principle. These fluctuations are usualy presented as the variance, or second central moment, since the average, or first moment, is zero.[19] The variance is always positive, as required, and its magnitude is usually taken as an inverse measure of goodness. That is, we want the variance of a measured or calculated variable to be as small as possible. It seems entirely reasonable that the equilibrium system would have the smallest variance.

For example, the variance of the local energy, $E_{\text{loc}} = H\psi/\psi$, has long been used to gauge the goodness of an approximate wave function, ψ.[20] The variance, $(\overline{E^2_{\text{loc}}} - \bar{E}^2_{\text{loc}})$, will approach zero as the true wave function is approached, since then the local energy will be a constant equal to the exact energy. Note that the exact energy will also be a minimum compared with the other $\bar{E}_{\text{loc}}$.

The energy of an electromagnetic field, confined to a hohlraum, is given by

$$E = \frac{1}{8\pi}(\overline{F^2} + \overline{B^2}) \tag{4.17}$$

where F is the electric field strength, and B is the magnetic field. Since $\bar{F} = \bar{B} = 0$, the minimum energy is where the variance of F and B is minimum. In Brownian motion, the mean values of $\overline{p^2}$ and $\bar{p}$, the momemtum, are given by

$$\overline{p^2} = 3MkT \qquad \text{and} \qquad \bar{p} = 0 \tag{4.18}$$

where M is the molecular weight. The smallest value of the variance again goes with the minimum value of the energy.

New information can be obtained by applying the principle of minimum variance. The fluctuations of the energy are given by[12]

$$C_v T = \beta \langle (E - \langle E \rangle)^2 \rangle \tag{4.19}$$

If T is constant, we conclude that the heat capacity of the equilibrium state is a minimum, compared with nearby non-equilibrium states. Clearly in the latter states, if there is excess energy, there will be a greater spread of energy values.

A larger value of C_v automatically follows from Equation (4.19). A convenient way to generate a non-equilibrium state at T is to imagine it to be the equilibrium state at a slightly higher temperature. The total energy always increases with increasing temperature. But the heat capacity, C_v, also always increases with increasing temperature. Hence the equilibrium state has the lowest C_v, since it has the properties of the lowest temperature.

TESTS OF THE PMH

As mentioned earlier, the restriction to constant μ and v, if absolute, would be very severe. As chemists we normally wish to examine changes in the nuclear positions, leading to chemical change. Fortunately, there are cases where this can be done, keeping μ and v fixed.[21] Start with a molecule in its equilibrium geometry and calculate its obital energies at, or near, the HF level. Now distort the molecule a small distance along directions given by the vibrational symmetry coordinates and re-calculate the orbital energies. By using the full set of symmetry coordinates, the hardness can be probed for all possible changes in the equilibrium geometry.

Table 4.3 shows the results of such calculations for the CO_2 molecule.[22] In order to understand the results, Figure 4.4 shows the normal modes of vibration of CO_2. The asymmetric modes, Π_u and Σ_u, differ from the symmetric mode, Σ_g, in several respects. First of all, they destroy some elements of symmetry, changing the point group to C_{2v} and $C_{\infty v}$, respectively.

Secondly, positive deviations from equilibrium produce a configuration which gives the same average nuclear potential as negative deviations. Therefore, ε_{HOMO}, ε_{LUMO}, μ and η must be the same for both. If we let Q represent a symmetry coordinate, then both $(\partial\mu/\partial Q)$ and $(\partial\eta/\partial Q)$ must be equal to 0 at the equilibrium Q_0.

Table 4.3. Distortions of Carbon Dioxide from the Equilibrium Geometry[a]

R_1 [Å]	R_2 [Å]	θ [deg]	$-\varepsilon_{HOMO}$	ε_{LUMO}	-2μ [a.u]	2η [a.u.]
1.1352	1.1352[b]	180[b]	0.5423	0.2307	0.3116	0.7730
1.1052	1.1652	180	0.5405	0.2313	0.3092	0.7718
1.1652	1.1052	180	0.5405	0.2313	0.3092	0.7718
1.1052	1.1052	180	0.5438	0.2391	0.3047	0.7829
1.1652	1.1652	180	0.5407	0.2066	0.3341	0.7473
1.1352	1.1352	175	0.5419	0.2098	0.3321	0.7517
1.1352	1.1352	185	0.5419	1.2098	0.3321	0.7517

[a] From Reference 22.
[b] Equilibrium values.

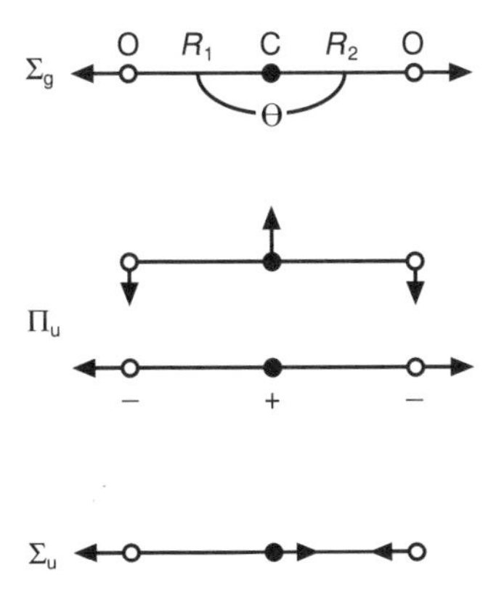

Figure 4.4 Normal vibrations of CO_2 in the D_{2h} point group.

Also, if we expand the energy as a power series in ΔQ, the linear term must vanish, and the quadratic term is the first non-vanishing one. Symmetry arguments can be used to show that $(\partial v_n/\partial Q)$ and $(\partial v_e/\partial Q)$ are separately equal to 0 when averaged.[17] Here v_n and v_e refer to the potentials of the nuclear repulsion and the nuclear–electron attraction, respectively. Hence, for the non-totally symmetric distortions, we have met the restrictions of Parr and Chattaraj. Accordingly, the hardness should be a maximum at Q_0. Table 4.3 shows that this is the case. Note that μ can either decrease or increase upon distortion from equilibrium.

The totally symmetric mode gives different results. The hardness and μ both increase steadily as the nuclei approach each other. If the nuclei coalesced, μ would be a maximum. This does not happen because at Q_0 we have the condition

$$\langle(\partial v_e/\partial Q) + (\partial v_n/\partial Q)\rangle = 0 \tag{4.20}$$

Thus the equilibrium value of Q is determined by the Hellman–Feynman theorem of balanced forces, and not by the maximum value of η. This is not a violation of the PMH since neither v nor μ remains constant.

These results for CO_2 are typical. The same features have been found for H_2O,[23] NH_3[21] and C_2H_6.[21] The last of these molecules is a good test, since there are three symmetric modes and nine asymmetric ones. Whereas the equilibrium distances and angles in a molecule are determined by Equation (4.20), the existence of symmetry in a molecule is determined by the hardness. Similar calculations have now been made for many molecules.[24]

If the hardness decreases upon any distortion that destroys an element of symmetry, that element is stable. If η increases, the molecule will spontaneously distort and the element will vanish. An example would be NH_3 in a planar form.[25] These results are strikingly similar to those deduced from the second-order Jahn–Teller effect.[26] Symmetry arguments can be used to predict the correct point group of a molecule, but not the equilibrium bond angles or bond distances when these are changed by symmetric distortions.

Datta was the first to point out an interesting corollary of the PMH. The transition state of a chemical reaction, which has a maximum energy, by definition, should have a minimum hardness as well.[25] Figure 4.5 shows the variation of both the total energy and the hardness for the inversion reaction of NH_3. The reaction coordinate is the angle between an N–H bond and the three-fold axis, so that 90° is the planar form. It is also the transition state (TS) for the inversion reaction of ammonia. The reaction coordinate has A_2'' symmetry at the TS, but is A_1 thereafter. The double potential well is at $\alpha = 65°$ and 115°. The hardness is clearly a minimum at $\alpha = 90°$, as expected, just as the energy is a local maximum. The energy in this case is the total energy. But symmetry arguments show that, for a non-symmetric reaction mode, all quantities such as V_{nn}, E_{el} and E must be extrema at the TS.[26] Since V_{nn}, the nuclear repulsion energy, is a minimum at this point, E_{el} must be a maximum.

That the hardness would be a minimum at the TS is expected on other grounds. Long ago Bader pointed out that an activated complex which exists at the TS must have low-lying excited states, or a very small HOMO–LUMO gap.[27] This follows from the second-order Jahn–Teller effect since activated complexes,

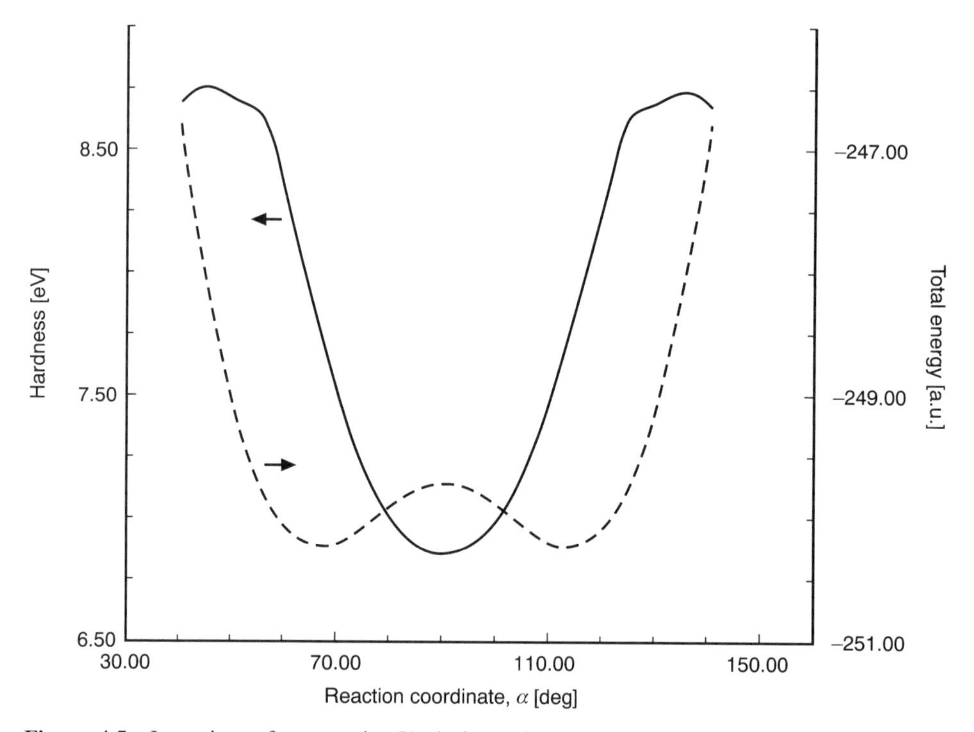

Figure 4.5 Inversion of ammonia. Variation of the total energy (broken line) and the hardness (solid line) as the angle, α, is changed. Hardness is in eV, and total energy in au. From Reference 25.

by definition, are extremely reactive. A formal proof of the Maximum Softness Principle has been supplied, in a sense, by Parr and Gásquez.[28]

These workers defined a quantity, the hardness functional $\mathscr{H}$, which had two properties of interest. For ground states, $-\mathscr{H}$ became equal to the grand potential, $E - N\mu$. Also $\mathscr{H}$, while not equal to the global hardness, η, behaves like it with respect to small changes. Thus we have

$$\Delta(E - N\mu) = -\Delta\mathscr{H} \tag{4.21}$$

This says that, if μ is constant, as $E = E_{el}$ decreases, the hardness will increase, and vice versa. A maximum in E means a minimum in $\mathscr{H}$, and a minimum in E means a maximum in $\mathscr{H}$: this is the PMH. Any state that is an extremum for both E_{el} and V_{nn} will be an extremum for the hardness.

The correctness of these conclusions is borne out by a number of studies made on the variations of μ and E in rotational isomerization reactions.[29] These have the interesting feature of having double-well potential energies with an intervening barrier, or a single well with two barriers.

As an example of the latter case, consider H_2O_2. This molecule is unstable in both the *cis* and *trans* forms, and is stable at $\alpha = 70°$ setting $\alpha = 0°$ for the *trans* form. Figure 4.6 shows the calculated variations of both the potential energy and the hardness as a function of α.[30] The mirror-image relationship between the two is what is predicted by Equation (4.21). At both 0° and 180°, η is a minimum and V is a maximum. This is required since the reaction coordinate is asymmetric at these points. The point groups are C_{2h} for the *trans* isomer and C_{2v} for the *cis*. The equilibrium point group is C_2, at $\alpha = 60°$, somewhat off the experimental value of 70°.

In all these applications of the grand potential, the energy which appears in Equation (4.5) is the electronic energy. The nuclear–nuclear repulsion energy does not appear explicitly, since it is supposedly constant. Also, the proof for the PMH is valid for small departures from equilibrium only. Therefore it cannot be assumed that the relative stabilities of isomers can be linked to the relative sizes of the HOMO–LUMO gap.

For example, some results for several possible structures of Si_4, calculated by DFT methods, are given in Table 4.4.[31] The most stable form is a rhombus, or diamond shape. This also has the greatest hardness. All three structures are readily interconvertible by small distortions from the stable structure.

Compare this case with that of possible structures for the (unknown) molecule P_6. Warren and Gimarc have made accurate SCF MO calculations on five of the possible valence isomers.[32] These were selected by looking at the five most stable valence isomers of C_6H_6, since P_6 is valence iso-electronic with C_6H_6. Replace each CH unit by a P atom in the structures **1**–**5** in Table 4.5. The energy values shown are for the P_6 isomers. There is no relationship between the relative energies and either the gap size or the chemical potential. The most stable isomer has the benzvalene structure, and almost the smallest gap.

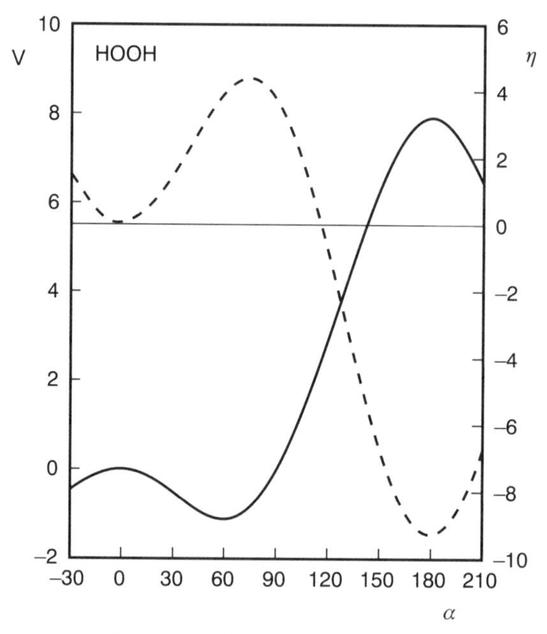

Figure 4.6 Calculated variation for H_2O_2 of the potential energy, V (solid line), and the hardness, η (broken line), as a function of the angle of rotation, α. The *trans* form is taken as the reference so that α, V and η are all zero. Energies are in kcal/mol. Reproduced with permission from G.I. Cárdenas-Jirón and A. Toro-Labbé, *J. Phys. Chem.*, **99**, 5325 (1995). © 1995 American Chemical Society.

In this example, the interconversion of isomers requires more extensive nuclear movement. What the PMH does predict is that each isomer is a local minimum on the energy–nuclear position hypersurface. The hardness will be a maximum with respect to all asymmetric distortions from the geometry of the local energy minimum. These predictions have been verified by accurate calculations on all five P_6 isomers.[33] As expected, for symmetric distortions, no maxima are found.

All of the previous examples have been based on the approximation

$$\eta = (\varepsilon_{\mathrm{LUMO}} - \varepsilon_{\mathrm{HOMO}})/2 \tag{4.22}$$

Table 4.4 Calculated Energy Values for Possible Si_4 Structures [eV]

	Structure		
	D_{2h}	D_{4h}	T_d
Total energy [eV]	0.00	1.98	2.27
η [eV]	1.67	1.52	1.50
$-\mu$ [eV]	4.57	3.70	5.05

Table 4.5 Calculated Energy Values for Possible P_6 Structures

	Structure				
	Benzene 1	Prismane 2	Dewar benzene 3	Benzvalene 4	Bicyclopropenyl 5
Energy [kcal/mol]	30.9	6.50	18.0	0.0	20.6
η [eV]	4.08	4.57	4.22	4.02	4.01
$-\mu$ [eV]	5.60	4.52	5.02	4.75	4.29

The orbital energies, ε, also have been calculated by methods of quite variable sophistication and accuracy. Yet the results always seem to support the concept that "molecules tend to be as hard as possible". Even a method as primitive as Hückel MO theory can be used with good results.[12]

A study has been made of the effect of basis set selection on *ab-initio* SCF calculations of molecular hardness.[34] Not surprisingly, the value of the LUMO energy, which is set equal to $-A$, is the term which changes most. This means that the hardness calculated from Equation (4.22) will be very sensitive to the quality of the calculations. Of course, μ will also vary greatly. But as long as the same basis set, or its equivalent, is used to compare different structures, or similar molecules, relative values of η and μ are found which are usually reliable.

A test of the PMH which is quite different, and more general, has recently been given.[35] It follows earlier work by Gyftopoulos and Hatsopoulos, who used a grand canonical ensemble with a limited number of discrete energy levels, so that the distribution function was known.[36] Those authors then calculated the electronic chemical potential, μ, which was found to be $\mu^0 = (I + A)/2$. The ensemble was a collection of systems containing the three species M^0, M^+, M^- and with energy levels E°, $(E^\circ + I)$ and $(E^\circ - A)$.

The ensemble average electronic energy can then be calculated in the usual way, and the softness from Equation (4.7):

$$\langle E \rangle - E^\circ = \frac{2\eta^\circ \exp(-\beta\eta^\circ)}{1 + 2\exp(-\beta\eta^\circ)} \tag{4.23}$$

$$\langle \sigma \rangle / \beta = \frac{2\exp(-\beta\eta^\circ)}{1 + 2\exp(-\beta\eta^\circ)} \tag{4.24}$$

These are the equilibrium values. The softness $\langle\sigma\rangle$ is not the same as $\sigma^\circ = 1/\eta^\circ$ where $\eta^\circ = (I - A)/2$. It may be thought of as an additional softness due to the finite temperature. At zero Kelvin, $\langle\sigma\rangle = 0$.

For two different equilibrium systems, Equations (4.23) and (4.24) tell us at once that the system of higher energy has the greater softness. The restrictions are constant temperature and constant $\langle N\rangle = N^\circ$. The restriction on μ° has been removed. This proof of the PMH would apply to small changes in the internuclear distances, since each set of nuclear positions corresponds to an equilibrium value of the electronic energy.

A more general case is that of a non-equilibrium distribution that is close to an equilibrium distribution. The parameters μ, η, E, I, v and A are all allowed to change, but $\langle N\rangle = N^\circ$ is maintained. It is then found that $\bar{\sigma} - \langle\sigma\rangle$ increases as $\bar{E} - \langle E\rangle$ increases, where the overbars indicate averages over the non-equilibrium ensemble. An interesting restriction is that $\Delta A > \Delta I$, i.e., the energy of the LUMO must change more than the energy of the HOMO. This is usually the cases for non-equilibrium electron distributions, as well as non-equilibrium geometries.

Adding more energy levels, such as the energies of M^{2+} and M^{2-}, has no practical effect on these results. However, they do depend on ignoring the contribution of TS in Equation (4.5). But the entropy in question is only the entropy associated with the different electronic degeneracies of M^0, M^+, M^-. At any reasonable temperature, the changes in TS can be ignored. The model calculation then strongly supports the PMH, as derived by Parr and Chattaraj on very different grounds.[13] It also supports the essential requirement that the non-equilibrium ensembles have an energy higher than equilibrium, and that small variations in μ and v are permissible.

THE SOFTNESS OF EXCITED STATES

A finite-temperature DFT is, in effect, an excited-state theory. As the temperature increases, more and more excited states are mixed in with the ground state. The excited states of greatest chemical importance are the electronic ones, including the ionized states M^+ and M^-. Excited vibrational states will also play a role since the equilibrium electron density function, which determines all the properties, depends on the mix of electron density functions for all the excited states, as well as the ground state.

The problem is that the Hohenberg–Kohn theorem is strictly for ground states. Therefore we cannot say, in general, that the energy of an excited state is a unique functional of its charge density. The main stumbling block is that the wave function for any excited state must be orthogonal to the wave functions for all lower states. A case for which this difficulty disappears is when the excited state is the lowest state of a given symmetry species.[37] Then the energy is determined by ρ for that state.

The search for a workable DFT of excited states is an important and active field at the present time.[38] There has been little done, however, on the concepts of chemical potential and hardness, which is our main concern. If we assume that Equation (4.1) is valid, we can draw some conclusions about μ and η for excited states.

Figure 4.7 displays the MO–energy diagrams for a ground-state molecule M^0, its first excited state, M^*, and the states which give the minimum I and the maximum A, M^+ and M^-. These are the same whether we start with either the ground or excited state. But the energy required to go from M^0 to M^+ is clearly larger than that for M^* to M^+. The difference is just the excitation energy, E^*. Similarly the energy gained in going from M^0 to M^- is less than that for M^* to M^- by the same amount, E^*. We have

$$(I^* + A^*)/2 = (I^\circ + A^\circ)/2 = -\mu^{*-} = -\mu^\circ \tag{4.25}$$

$$(I^* - A^*)/2 = (I^\circ - A^\circ)/2 - E^* = \eta^* \tag{4.26}$$

There is no change in the electronic chemical potential, but the hardness of the excited state is less than that of the ground state.

This conclusion can be drawn in another way. Write the reactions

$$2M^0 = M^+ + M^- \qquad \Delta E = (I - A) = 2\eta \tag{4.27}$$

$$2M^* = M^+ + M^- \qquad \Delta E = (I - A) - 2E^* \tag{4.28}$$

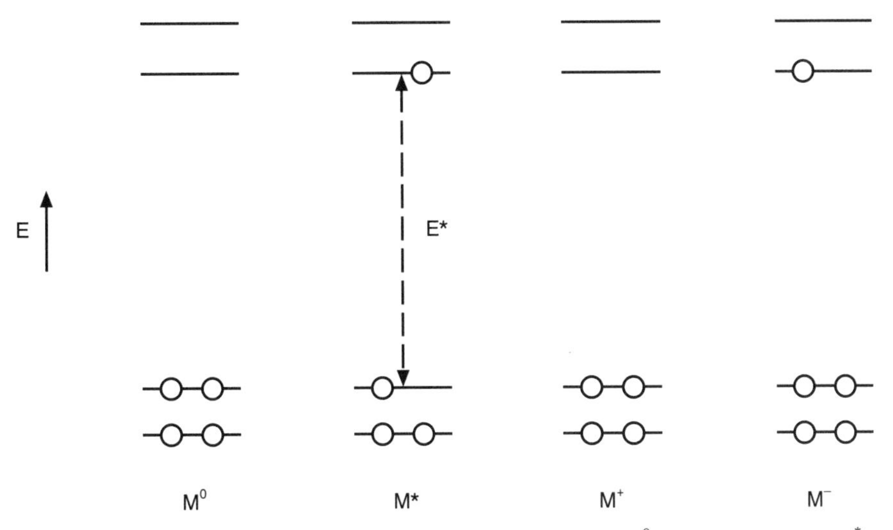

Figure 4.7 MO energy diagrams for a ground-state molecule, M^0, an excited state, M^*, and the lowest-energy M^+ and M^-.

Reaction (4.27) is sometimes used to define hardness.[39] Then reaction (4.28) shows that excited-state molecules are softer than ground-state ones. Furthermore, it does not matter what kind of excess energy M* has. It can be electronic, vibrational or even rotational and translational. Any energy above the ground-state energy at 0-K makes it easier to distort the electron cloud.

This result is in agreement with our previous demonstrations that non-equilibrium ensembles are softer than equilibrium ones, provided they have excess energy. It also agrees with the results of the three-level model, as contained in Equations (4.23) and (4.24). The gain in softness is directly proportional to the gain in energy, the proportionality constant being β/η°. As the temperature increases, both the energy and the softness increase, though changes can occur at very high temperatures, over 2000 K.[40]

Another general criterion for softness is increased chemical reactivity. Certainly excited-state, or energetic, molecules are more reactive than ground-state molecules. The more excess energy they have, the more reactive they are. What other properties of excited-state molecules may be related to increased softness? A soft molecule, or atom, is usually thought of as one with a more diffuse, or less compact, electron cloud than a similar hard molecule or atom.

The electron-cloud picture of an atom is often considered as a consequence of the Uncertainty Principle. This suggests that the compactness of the cloud should be considered in six-dimensional phase space, rather than in the three dimensions of ordinary space. Then the uncertainty of a specific state of an atom, or molecule, can be expressed as $\Delta p_x \Delta x$ for each degree of freedom. The conventional way to calculate the uncertainty is by way of the variance of p_x and x[41]:

$$\Delta p_x \Delta x = [\langle (p_x^2 - \langle p_x \rangle^2) \rangle \langle (x^2 - \langle x \rangle^2) \rangle]^{1/2} \tag{4.29}$$

Now we can readily find the uncertainty of the various translational, rotational and vibrational energy levels that a simple molecule would have. The electronic energy is more complicated, but the result for a hydrogen atom may be taken as representative. Except for the latter, a diatomic molecule gives the formulae shown in Table 4.6. The rotational and electronic terms have been averaged over m_l to give spherical symmetry.

The important result is that the uncertainty increases steadily with the relevant quantum numbers. This agrees with the expectation that excited states are softer than the ground state. The electron cloud, or particle density, is less compact in phase space for the excited levels in all four cases.

Table 4.6 has another interesting interpretation. The uncertainty in a given system is used as a measure of its entropy:[42]

$$S_{\mathrm{I}} = R \ln \frac{(\Delta p_x \Delta x)}{h/2\pi} \tag{4.30}$$

Table 4.6 Uncertainty of the Various Energy Levels of a Diatomic Molecule

Energy	$\Delta p_x \Delta x$
Translational	$n_x\left(\frac{\pi^2}{12}-\frac{1}{2n_x^2}\right)^{1/2} h/2\pi$
Rotational	$(J(J+1))^{1/2}h/2\pi$
Vibrational	$(2v+1)^{1/2}h/4\pi$
Electronic	$\frac{n}{3}\left[1+\frac{3}{2}\left\{1-\frac{\ell(\ell+1)-1/3}{n^2}\right\}\right]^{1/2} h/2\pi$

Table 4.7 Entropy of H_2X Molecules

	$H_2O(g)$	$H_2S(g)$	$H_2Se(g)$
S° [J/mol K]	188.8	205.8	219.0

Written in this way S_I is called the information, or uncertainty, entropy. But in fact, using the results of Table 4.6, and calculating suitable average quantum numbers, the ordinary thermodynamic entropy can be obtained from Equation (4.30).

Thus as the temperature increases, the energy, the entropy, the uncertainty and the softness all increase. There is other evidence showing a relation between softness and entropy. For example, in Table 4.2 we found that the softness always decreased as molecules were formed from atoms or radicals. But the entropy also decreases, as translational motion is converted to rotational and vibrational. Also, if we examine a series of related molecules, we will find that the entropy increases as the softness increases (Table 4.7).

In spite of all this, there is an important counter-example. The transition state for a chemical reaction is very soft, as we have seen, and is higher in energy than the reactants. Nevertheless, the entropy of activation, $\Delta S^\ddagger$, is negative more often than not. Hence the entropy of the activated complex is less than the entropy of the reactants.

HARDNESS AND THE ELECTRONIC ENERGY

It would be very useful if a quantitative relationship could be found between hardness and the energy of chemical systems. Both μ and η are small numbers compared with the total energies of such system. Quantities of chemical interest, such as bond energies and activation energies, are calculated as small differences between very large numbers. This greatly magnifies the errors of calculation. If, instead, we could calculate changes in μ or η and convert them to changes in energy, in principle greater accuracy could be achieved.

The preceding sections have shown that it is η, and not μ, which is more closely related to energy. Also, the relationship is an inverse one in that $+\Delta E$ means $-\Delta\eta$. An annoying restriction is that μ should be constant, or change very little. Also it is the electronic energy, E_{el}, which appears in DFT-based equations such as (4.5) and (4.21), where N is the number of electrons. Changes in nuclear–nuclear repulsion are usually easy to calculate, however.

Changes in μ can be accommodated, if we assume that it is the grand potential, Ω, which is related to the hardness. At low temperature,

$$\Omega = E_{el} - N\mu = f(\eta) \tag{4.31}$$

Let us examine Equation (4.31) for the case of fundamental chemical importance, the bringing together of two atoms to form a molecule. The simplest example is the formation of H_2^+ from a hydrogen atom and a proton. The electronic energies of $H_2^{2+}(E_{el} = 0)$, H_2^+ and H_2 are all known for every value of R, the internuclear separation, from zero to infinity:

$$\underset{R=\infty}{H + H^+} \qquad H_2^+ \qquad \underset{R=0}{He^+} \tag{4.32}$$

We can easily find I and A and hence μ and η. The results are given in Table 4.8. There is a simple relationship between these quantities, which we can write as

$$(E_{el} - N\mu) = -N^2\eta \tag{4.33}$$

The reason for including N and N^2, both equal to one, is that the exact definitions of η and η demand them, or their dimensional equivalent

$$\mu = (\partial E/\partial N)_v; \qquad 2\eta = (\partial^2 E/\partial N^2)_v \tag{4.34}$$

If we try to apply Equation (4.33) to other diatomic molecules or to atoms, we find that it does not work, except for H_2 and probably He. Hartree–Fock calculations have been made on H_2, Li_2, N_2[43] and HF,[44] so that E_{el}, μ and η

Table 4.8 E_{el}, μ and η Values for H_2^+

R [a.u]	$-E_{el}$ [a.u]	$-\mu$ [a.u.]	η [a.u.]
∞	0.500	0.500	0.000
2.00	1.103	0.819	0.284
0	2.000	1.452	0.548

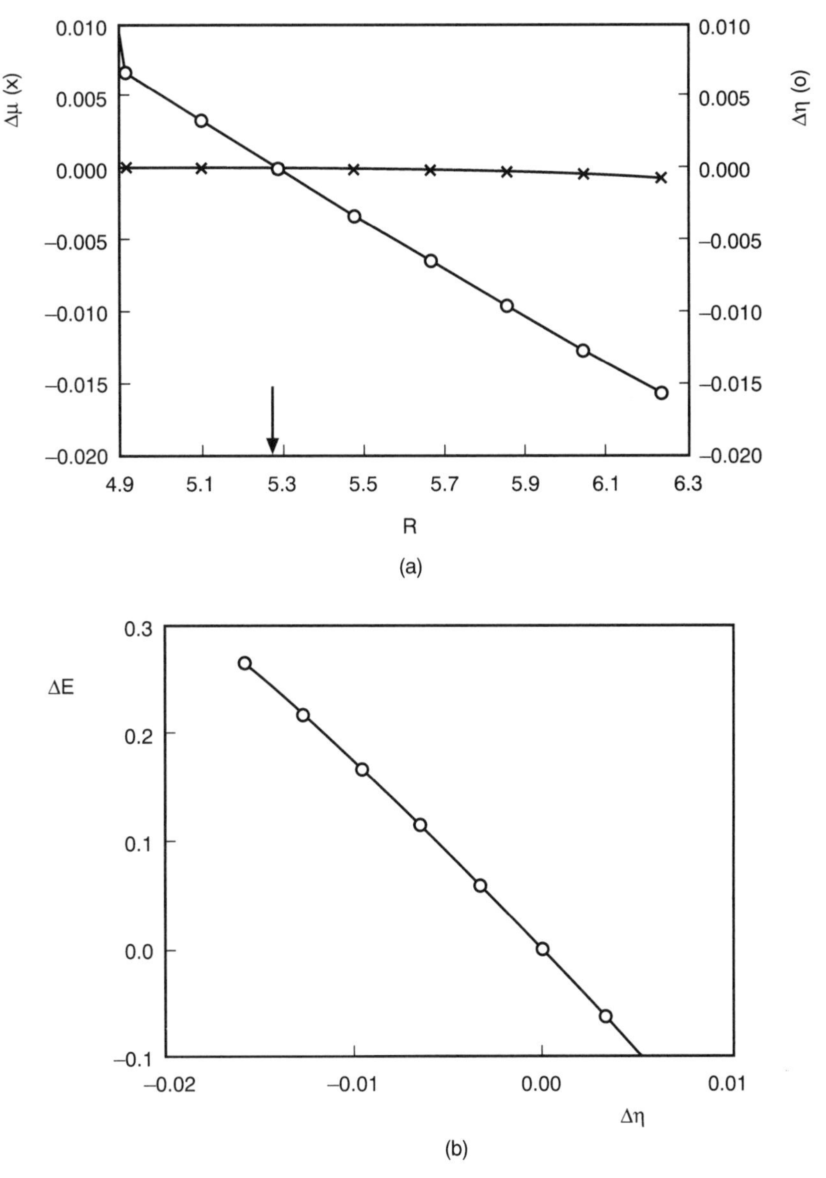

Figure 4.8 (a) Changes in μ and in η for Li_2 as the internuclear distance, R is changed from R_0, indicated by the arrow. (b) Changes in the electronic energy, E, plotted against changes in η. Energies are in a.u. Reproduced with permission from J.L. Gasquez *et al.*, J. Phys. Chem., *97*, 4059 (1993). © 1993 American Chemical Society.

are known at several values of R in the region of the equilibrium value, R_0. Figure 4.8 shows some results for Li_2, which may be taken as fairly typical.

Figure 8(a) shows μ and η as a function of R, in terms of changes from the values at R_0. Figure 8(b) has $\Delta\eta$ plotted against the change in the electronic energy. While $\Delta\mu$ is small, $\Delta\eta$ is a linear function of ΔE_{el} with a negative slope, as Equation (4.33) would predict, but the magnitude of the slope is wrong by a factor of about four. The same results are found for H_2, N_2 and HF, except that the slopes are variable, being either more or less than N^2, except for hydrogen.

Going back to equation (4.21),

$$\Delta(E - N\mu) = -\Delta\mathscr{H} \tag{4.21}$$

we see that the slope depends on the changes in the hardness functional, $\mathscr{H}$. The definition of $\mathscr{H}$ is given by[28]

$$\mathscr{H}[\rho] = \langle(\delta F[\rho]/\delta\rho)\rangle - F[\rho] \tag{4.35}$$

where $F[\rho]$ is the universal functional of DFT for the kinetic energy and the electron-electron repulsion energy. Clearly $\mathscr{H}$ for H, H_2^+ and H_2 is just the hardness, η, of Equation (4.33) times N^2. In other cases it is proportional to η, at least for a certain range of variation.

Unfortunately, we do not yet know how to calculate $\mathscr{H}$ with any accuracy, so that the constants of proportionality are not known. But this is an area where progress can be made, hopefully, by further study of Equation (4.35). Another problem also exists, in attempting to relate changes in η to changes in energy. The values of μ and η depend very much on the quality of the method used to calculate them. While the relative values that have been calculated may be quite good, as mentioned earlier, absolute values are not. The calculation of electron affinities is especially difficult. On the positive side, if we know R_0, then we also know $(\partial E_{el}/\partial R)$ without any further quantum mechanical calculations. It is simply equal to the force due to nuclear repulsion, $Z_A Z_B/R_0^2$.

The nuclear–nuclear repulsions must be taken into account; often, but not always, they account for relative stabilities. It is worthwhile looking at all the relevant parameters in a few cases, taking theoretical results. Table 4.9 shows the results for two pairs of isomers: naphthalene and azulene, $N = 68$; and cyanic acid and isocyanic acid, $N = 14$. The most stable isomers, naphthalene and

Table 4.9 Calculated Parameters for Pairs of Isomers

	$-E$ [a.u.]	V_{nn} [a.u.]	$-E_{el}$ [a.u.]	$-N\mu$ [a.u.]	η [eV]
Naphthalene[(a)]	382.788	459.750	842.538	9.003	5.70
Azulene[(a)]	382.708	455.169	837.877	9.629	4.41
HCN[(b)]	92.877	24.311	117.188	2.056	9.51
HCN[(b)]	92.859	24.503	117.362	1.892	9.34

[(a)] R.J. Buenker and S.D. Peyerimhoff, Chem. Phys. Lett., **3**, 37 (1969).
[(b)] Reference 46.

HCN, cannot be predicted by V_{nn}, E_{el}, $N\mu$ or Ω. The only consistency is the size of the HOMO–LUMO gap, used to calculate η.

Similar results are found in detailed calculations on ten pairs of isomers of the type HAB and HBA.[46] However, in only eight of the ten cases does the most stable isomer have the greater hardness. In any event, the PMH does *not* predict that the hardness should correlate with the total energy, but with the electronic energy, or Ω. Of course, a larger gap going with greater stability is a reasonable result, though not predicted.

The failure in three cases to correlate E_{el} with the energy gap means that going from HCN to HNC, for example, is too large a change to meet the requirement of PMH. It is found in all cases that the two isomers, in their most stable geometry, are local maxima in hardness, as predicted. Also the transition state for interconverting HAB and HBA is always much softer than either isomer. Apparently, going from one stable structure to the TS is a smaller change than going all the way to the other isomer, and the PMH holds.

REFERENCES

1. R.G. Pearson, *J. Chem. Ed.*, **64**, 561 (1987).
2. L.S. Bartell, *J. Chem. Ed.*, **45**, 754 (1968).
3. E.L. Muetterties and B.F. Beier, *Bull. Soc. Chim. Belg.*, **84**, 397 (1979).
4. R.G. Pearson, *Acc. Chem. Res.*, **26**, 250 (1993).
5. R.G. Pearson, *Inorg. Chim. Acta*, **198–200**, 781 (1992).
6. J.P. Perdew, R.G. Parr, M. Levy and J.L. Balduz, Jr., *Phys. Rev. Lett.*, **49**, 1691 (1982).
7. R.G. Parr and Z. Zhou, *Acc. Chem. Res.*, **26**, 256 (1993).
8. J. Robles and L.J. Bartolotti, *J. Am. Chem. Soc.*, **106**, 3723 (1984).
9. A. Streitwieser, Jr., *Molecular Orbital Theory for Organic Chemists*, John Wiley and Sons, New York, 1962, p. 290.
10. D.M.P. Mingos, *Acc. Chem. Res.*, **17**, 311 (1984).
11. M.G. Mayer and J.H.D. Jensen, *Elementary Theory of Shell Structure*, John Wiley and Sons, New York, 1955, p. 58.
12. Z. Zhou and R.G. Parr, *J. Am. Chem. Soc.*, **112**, 5720 (1990).
13. R.G. Parr and P.K. Chattaraj, *J. Am. Chem. Soc.*, **113**, 1854 (1991).
14. W. Wang and R.G. Parr, *Proc. Natl. Acad. Sci. USA*, **82**, 6723 (1985).
15. For general discussions see (a) D. Chandler, *Introduction to Modern Statistical Mechanics*, Oxford University Press, New York, 1987; (b) J. Keizer, *Statistical Thermodynamics of Nonequilibrium Processes*, Springer-Verlag, New York, 1987.
16. H. Nyquist, *Phys. Rev.*, **32**, 110 (1928).
17. L. Onsager, *Phys. Rev.*, **37**, 4053; idem, ibid., **38**, 2265 (1931).
18. H.B. Callen and T.A. Welton, *Phys. Rev.*, **83**, 34 (1951).
19. For an elementary discussion of fluctuations, see D.A. McQuarrie, *Statistical Mechanics*, Harper and Row, New York, 1973, Chapter 5.
20. A.A. Frost and R.E. Kellogg, *Rev. Mod. Phys.*, **32**, 313 (1960); R.G. Pearson and W.E. Palke, *Int. J. Quantum Chem.*, **37**, 103 (1990).

21. R.G. Pearson and W.E. Palke, *J. Phys. Chem.*, **96**, 3283 (1992).
22. W.E. Palke, unpublished calculations.
23. S. Pal, N. Vaval and R. Roy, *J. Phys. Chem.*, **97**, 4404 (1993).
24. P.K. Chattaraj, S. Nath and A.B. Sannigrahi, *Chem. Phys. Lett.*, **212**, 223 (1993).
25. D. Datta, *J. Phys. Chem.*, **96**, 2409 (1992).
26. R.G. Pearson, *Symmetry Rules for Chemical Reactions*, Wiley–Interscience, New York, 1976, Chapters 1 and 3. See also G. Makov, *J. Phys. Chem.*, **99**, 9337 (1995).
27. R.F.W. Bader, *Can. J. Chem.*, **40**, 1164 (1962).
28. R.G. Parr and J.L. Gázquez, *J. Phys. Chem.*, **97**, 3939 (1993).
29. P.K. Chattaraj, S. Nath and A.B. Sannigrahi, *J. Phys. Chem.*, **98**, 9143 (1994); G.J. Cárdenas-Jirón, J. Laksen and A. Toro-Labbé, *J. Phys. Chem.*, **99**, 5325 (1995).
30. G.I. Cárdenas-Jirón and A. Toro-Labbé, *J. Phys. Chem.*, **99**, 12730 (1995).
31. M. Galván, A. Dal Pino, Jr. and J.D. Joannopoulos, *Phys. Rev. Lett.*, **70**, 21 (1993).
32. D.S. Warren and B.M. Gimarc, *J. Am. Chem. Soc.*, **114**, 5378 (1992).
33. D.S. Warren and B.M. Gimarc, *Int. J. Quantum Chem.*, **49**, 207 (1994).
34. S. Nath, A.B. Sannigrahi and P.K. Chattaraj, *J. Mol. Struct. (Theochem.)*, **306**, 87 (1994).
35. P.K. Chattaraj, G.H. Liu and R.G. Parr, *Chem. Phys. Lett.*, **237**, 171 (1995).
36. E.P. Gyftopoulos and G.N. Hatsopoulos, *Proc. Nat. Acad. Sci. USA*, **60**, 786 (1965).
37. O. Gunnarson and B.I. Lundquist, *Phys. Rev. B*, **13**, 4274 (1976).
38. R.G. Parr and W. Yang, *Density Functional Theory for Atoms and Molecules*, Oxford Press, New York, 1989, pp. 204–208.
39. Reference 38, p. 96.
40. P.K. Chattaraj, A. Cedillo and R.G. Parr, *Chem. Phys.*, **204**, 429 (1996).
41. For example, see I.N. Levine, *Quantum Chemistry*, 2nd Ed., Allyn and Bacon, Boston, 1974, p. 63.
42. H.S. Robertson, *Statistical Thermophysics*, Prentice-Hall, Englewood Cliffs, NJ, 1993, Chapter 1.
43. J.L. Gazquez, A. Martinez and F. Mendez, *J. Phys. Chem.*, **97**, 4059 (1993).
44. S. Pal, R. Roy and A.K. Chandra, *J. Phys. Chem.*, **98**, 2314 (1994).
45. A.P. Hitchock, M. Tronc and A. Modelli, *J. Phys. Chem.*, **93**, 3068 (1989).
46. T. Kar and S. Schneiner, *J. Phys. Chem.*, **99**, 8121 (1995).

5 The Solid State

INTRODUCTION

There are several cogent reasons to include a chapter on the solid state in a treatise devoted to chemical hardness, and other concepts, derived from density functional theory. One is that DFT has been the theoretical method of choice in dealing with solid-state problems for a number of years.

This began with a landmark paper by Slater in which he proposed that the effects of exchange in the wave function could be replaced by an exchange potential, proportional to $\rho^{1/3}$ where ρ is the electron density function.[1] This followed earlier work by Dirac, in which he showed how to add the exchange energy to the Thomas–Fermi theory of the atom.[2] The exchange potential is also dependent on a factor, α, which is allowed to vary somewhat from its value in a uniform electron gas. The method is called the $X\alpha$ method and involves solving a series of one-electron wave equations in a self-consistent manner.

Slater was led to the $X\alpha$ method by the difficulty of using the Hartree–Fock method in the case of solids, which contain an enormous number of atoms and electrons. The HF method would require a complicated superposition of determinantal wave functions, and in addition was known to give some erroneous results.[3] The $X\alpha$ approach immediately showed its usefulness in calculating the physical properties of solids.[4]

At this point it should be noted that we are talking about crystalline solids where there is strong bonding between all nearest neighbors. That is, molecular solids are excluded. In the strong-bonding case, it is necessary to consider the entire crystal as one giant molecule. In such cases, DFT has great advantages over HF, as discussed in Chapter 2. In addition, very large molecules, such as proteins, can be handled in DFT by the "divide-and-conquer" method.[5] This technique breaks the molecule down into its subunits and solves each of these in a separate calculation. The results may then be joined together, obeying certain restrictions.

The solid-state equivalent of the divide-and-conquer method has been known for a long time. The natural subunits are the unit cells, which are repeated many times to form the macroscopic crystal. The wave equations need to be solved only for a single cell. The connection between the cells is made by multiplying by the factor exp $(ik \cdot R)$ when one goes from any point in one unit cell to the corresponding point in another unit cell. R is a vector marking the distance of the second cell from the first and k is a wave vector in reciprocal space. This procedure converts a local orbital, or wave function, to a Bloch function, named after its discoverer.[6] The periodicity of the lattice potential forces this kind of crystal wave function. An electron in such an orbital is then distributed over the entire crystal.

A second reason to consider solids here is that for this state the concept of physical hardness is important. Physical hardness is the resistance to a change in volume or shape of a solid object, produced by mechanical forces. Remembering that chemical hardness is subject to a restriction of constant nuclear positions, we see that physical hardness has the effect of removing this restriction. Nuclear positions must change, and this must be accompanied by a change in the electron density.

Finally, it turns out that the new definitions of electronic chemical potential, μ, and hardness, η, in DFT are actually old concepts in both solid-state physics and electrochemistry. In these fields it is often convenient to think of solids as havng electrons which are relatively free to move about, and which are independent components rather than appendages of the atoms.

Thus the equations,

$$\mu = (\partial E/N)_{S,V} \qquad \text{and} \qquad 2\eta = (\partial^2 E/\partial N^2)_{S,V} \tag{5.1}$$

where N is the number of electrons, are just simple thermodynamics. The second equation is not generally used, nor is the interpretation of chemical hardness. However the gap between the HOMO and the LUMO is an important one in solids. It is called the energy gap or the band gap.

BONDS IN SOLIDS

We will begin our study of solids by considering the chemical bonds that must exist between neighboring atoms. This will be done by the use of a semi-empirical MO theory, which is simple and gives much insight into the nature of the bonding.[6] It is a localized MO theory, even though the idea of a single crystal being a giant molecule calls for orbitals delocalized over the entire crystal.

These non-local orbitals are the canonical MOs of the giant molecule.[7] But we can always take linear combinations of the filled canonical MOs to produce an equivalent set of filled MOs which are localized. The elecron density and the energy are unchanged by this operation. The localized MOs are much easier to picture and to comprehend. Afterwards we will see how the delocalized MOs change the story.

The LCAO–MO model used here is very similar to models already developed for bonding in single molecules.[8] These are reasonably successful for predicting, or rationalizing, bond energies. Actually the LCAO–MO semi-empirical approach has been used in solid-state theory for some time.[9] It is called the tight-binding model. The emphasis has been on understanding the electrical and magnetic properties of solids, quite naturally. As a result these applications, while very useful in some ways, have not been good at predicting the cohesive energy of solids.[10]

From the chemist's point of view, this is probably the most important property of a solid. It is defined as the energy required to dissociate one mole of solid into its constituent atoms:

$$AB(s) = A(g) + B(g) \qquad \Delta E_{exp} \simeq \Delta H^{\circ} \tag{5.2}$$

We will write the cohesive energy as ΔE_{exp} since our immediate goal is to understand the experimental results already available from heats of formation of solids. We can ignore the small errors resulting from heats at 298 K rather than at absolute zero, the reference temperature for most theories.

The cohesive energy tells us about the strength of the chemical bonds in the solid. Its magnitude determines the stability and chemical reactivity of AB. Eventually it is the quantity which determines the structure of AB, since different possible structures will have different energies.

Another property of great interest is the ionicity of the bonding. To what extent do the atoms of the solid resemble neutral atoms, held together by covalent bonds, and to what extent are they like ions held together by electrostatic forces? This is a difficult question. Even if we have an accurate X-ray picture of the electron density of a compound, it is very hard to say whether atoms or ions are being shown. The same is true for an electron density calculated by accurate quantum mechanical methods.

Over the years numerous scales of "percent ionic character' have been proposed for both simple molecules and solids.[11] They may depend on some physical property, such as dipole moments or refractive indices, or be extracted from some theoretical analysis. The various scales usually agree on a rough ordering but the actual numbers vary widely. It seems to be impossible to define exactly what is meant by "percent ionic character". Therefore, the numbers in the various scales have meaning only in terms of the model from which they were derived. Each scale is useful for some experimental or theoretical property, but transfer to other properties is hazardous.

The model used here gives a different approach to the question of ionicity. The scale is based upon energy. The model can give a good approximation to the energy for completely covalent bonding and for completely ionic bonding. The relative values, ΔE_{cov} and ΔE_{ion}, give strong evidence for the nature of the bonding, especially if one or the other is quite close to he experimental ΔE_{exp}. The model also interpolates between the two limits in a way that is well grounded in fundamental theory. This then allows an estimate to be made of the ionicity that gives the maximum cohesive energy. The increase in this quantity from ΔE_{ion} or ΔE_{cov}, or ΔE_{mix}, is also important.

This method has only been possible for the alkali halides and the halides of group 11 (Cu, Ag, Au), until recently. We will apply it to all the binary compounds AB, often written as A^nB^{8-n}, where A is a metal and B a non-metal. The transition metals are not included, except for d^{10} and high-spin d^5: otherwise crystal field stabilization would have to be included.

The ionic model requires a knowledge of the higher electron affinities of atoms like S and P. Only the first affinity (A_1) is known for any of the elements. The second and third (A_2 and A_3) cannot be measured experimentally, and are nearly impossible to calculate theoretically.

However, very recently, a self-consistent set of values for A_2 (and A_3) has been proposed for O, N, S, P, Se and As atoms.[12] These come from a semi-empirical method using Slater orbitals, but calibrating the orbital exponents with the known values of I and A_1. Table 5.1 presents these new values in the form of the total charging energy, $\sum A_n$. This is the energy evolved in forming the ion B^{n-} from B. Estimates are also given for Te and Sb, but these cannot be found from Slater orbitals and are only crude extrapolations. The Prewitt–Shannon ionic radii R_B are listed for the various anions, and the orbital exponents, ζ, for the valence-shell Slater orbitals are also shown. These indicate that ions like P^{3-} and As^{3-} are more compact than might have been imagined.

The great majority of AB compounds have either a rock-salt structure (coordination number 6) or a sphalerite–wurtzite structure (coodination number 4). A good theory should predict the correct structure, or at least the correct coordination number (CN). Pauling had originally done this by means of his radius-ratio rules. However, it has been pointed out that these rules work only 50 percent of the time.[13] The evidence is now strong that CN6 compounds are highly ionic, and that CN4 comounds are more covalent.[14] The usual explanation for this is that CN6 has the higher Madelung constant, favoring

Table 5.1 Some Properties of the Monatomic Anions

Anion	$\sum A_n$ (a) [kcal/mol]	R_B (b) [Å]	ξ (c)	Anion	$\sum A_n$ (a) [kcal/mol]	R_B (b) [Å]	ξ (c)
F^-	78.4	1.19	2.32	Se^{2-}	−51	1.84	1.51
Cl^-	83.4	1.67	1.89	Te^{2-}	−45 (d)	2.07	–
Br^-	77.5	1.82	1.87	N^{3-}	−415	1.40 (e)	1.08
I^-	70.6	2.06	–	P^{3-}	−307	1.82 (e)	1.14
O^{2-}	−144	1.26	1.72	As^{3-}	−278	1.86 (e)	1.12
S^{2-}	−61	170	1.53	Sb^{3-}	−250 (d)	2.05 (e)	–

(a) Energy evolved in the process $B + ne^- = B^{n-}$
(b) Ionic radii or CN6, experimental.
(c) Orbital exponent for the valence-shell Slater orbital.
(d) Estimated only.
(e) Calculated, using data from ionic salts.

ionic bonding, and that CN4 compounds can form four tetrahedral hybrid orbitals at each atom, favoring covalent bonding. Some qualification is necessary for the latter explanation.

The accepted criteria for the strength of covalent bonds are due to Mulliken.[15] The exchange integral β increases with the ionization potentials (or electronegativities) of the bonded atoms. It also increases with the magnitude of the overlap integral S. Since a tetrahedral hybrid gives the maximum overlap with an orbital of a given neighboring atom, it is assumed that hybrid orbitals form the strongest bonds. But this is by no means true if there are several neighbors. In CH_4, for example, the total overlap is exactly the same for the symmetry-adapted MOs as for the four tetrahdral hybrids. All that hybridization does is to concentrate the orbital in a particular direction.

One might even believe that there is more covalent bonding with six nearest neighbors than with four, since there are more orbitals to overlap, but in extended systems such as the AB solids this is not true either. The total overlap is exactly the same for one s orbital and three p orbitals on each atom, whether the CN is 4 or 6. This results because the contribution of any AO to any MO is diluted according to the number of bonded atoms. This exactly compensates for the greater number of overlaps, providng the interatomic distances are the same. this is also true for CN8, as in the CsCl structure.

It is the distance feature that favors covalent bonding for CN4. As is well known, interatomic distances are greater for CN6 than for CN4, by about 5 percent, when the same atoms are bonded. This is a consequence of the greater number of nearest-neighbor repulsions. Since the overlap integral falls off exponentially with distance, it follows that β_{oct} is less than β_{tet}, leading to weaker covalent bonding.

DETAILS OF THE MODEL

For the case of completely ionic bonding, the Born–Mayer equation will be used:[16]

$$U = -\frac{MZ^2}{R} + mDe^{-R/\rho} \tag{5.3}$$

$$U_0 = -\frac{MZ^2}{R_0}(1 - \rho/R_0) \tag{5.4}$$

$$\frac{V_0}{\kappa} = \frac{MZ^2}{9R_0}(-2 + R_0/\rho) \tag{5.5}$$

Here m is the coordination number and M is the Madelung constant (1.748 for $m = 6$ and 1.64 for $m = 4$; no distinction is made between sphalerite and wurtzite structures).

The equilibrium interatomic distance R_0 is assumed to be known, in addition to the compressibility κ and the molar volume V_0. Actually, only a limited number of reliable values for κ are known,[17] but regularities in V_0/κ allow reasonable estimates to be made in other cases.[18] When Z is 2 or 3 small errors in these estimates can lead to large errors in the equilibrium potential energy U_0.

The equation for the cohesive energy becomes

$$\Delta E_{\mathrm{ion}} = -\sum I_n + \sum A_n - U_0 = \Delta E_{\mathrm{exp}} \tag{5.6}$$

For ΔE_{exp} the value of ΔH°_{298} is used.

The calculation for completely covalent bonding follows that of Coulson,[19] except for the way in which β is calibrated. No distinction is made between valence shell s- and p-electrons, and overlap integrals are ignored. Only interactions between nearest neighbors are considered. The unit taken is a single A atom and its nearest B neighbors. Each of these is shared by m A atoms, so the unit for which E is calculated is a single AB molecule.

Each atom forms four equivalent bonding MOs with its neighbors. The eight valence-shell electrons of the AB unit are put into these four MOs. Let N_{A} equal the average number of electrons on atom A and N_{B} the corresponding number on B, for each MO. Then we have[8,19]

$$N_{\mathrm{A}} + N_{\mathrm{B}} = 2 \tag{5.7}$$

$$E = 4[N_{\mathrm{A}}\alpha_{\mathrm{A}} + N_{\mathrm{B}}\alpha_{\mathrm{B}} + 2(N_{\mathrm{A}}N_{\mathrm{B}})^{1/2}\beta] \tag{5.8}$$

where α_{A} and α_{B} are the coulomb integrals.

As Coulson pointed out, completely covalent bonding in this case does not correspond to $N_{\mathrm{A}} = N_{\mathrm{B}} = 1$, but to $4N_{\mathrm{A}} = n$ and $4N_{\mathrm{B}} = 8 - n$. This gives zero net charge on atoms A and B, and α_{A} and α_{B} can be given their free-atom values. For reaction (5.1) (the dissociation of a solid into its atoms) there is then no change in the coulomb terms. The change in energy for Equation (5.1) becomes

$$\Delta E = -8\left[\left(\frac{n}{4}\right)\left(\frac{8-n}{4}\right)\right]^{1/2} \quad \beta = -2\beta[n(8-n)]^{1/2} \tag{5.9}$$

When $n = 1$, 2, 3 or 4 we have the 1–7, 2–6, 3–5 and 4–4 cases. Equation (5.9) gives $2\beta\sqrt{7}$, $2\beta\sqrt{12}$, $2\beta\sqrt{15}$ and 8β, respectively. If AB were a diatomic molecule with each atom contributing one electron to a single bond, we would have $\Delta E = 2\beta$.

For consistency, we add a repulsion term, as in Equation (5.2) According to Mulliken,[15] this term depends on the square of the overlap integral. Therefore the potential energy function for pure covalent bonding would be

$$U = -2C\mathrm{e}^{-R/2\rho} + mD\mathrm{e}^{-R/\rho} \tag{5.10}$$

$$U_0 = -C\mathrm{e}^{-R_0/2\rho} \qquad V_0/\kappa = U_0 R_0^2/18\rho^2 \tag{5.11}$$

The first term on the right-hand side in Equation (5.10) is the same as $-\Delta E$ in Equation (5.9), but explicitly showing that β depends on R. The dependence shown follows from the repulsion term, depending on the first power of the overlap integral, rather than on the square of it. Equation (5.10) is the solid-state version of the Morse equation.

The cohesive energy, ΔE_{cov}, is now equal to $-U_0$, which is equal to $\Delta E/2$ in Equation (5.9). Since β is to be fitted in any case, a knowledge of ρ is not so critical for covalent bonding. For orientation, ρ is equal to 0.305 Å for carbon and 0.486 Å for gray tin.

It is necessary to evaluate β in Equation (5.9) by fitting to some experimental data. One method has been to use the Mulliken criteria relating β to I and S. This was first done by Wolfsberg and Helmholz.[20] It requires separate consideration of s and p orbitals, however.

The best overall results are obtained by calibrating with single-bond energies for the elements, and using the arithmetic mean:

$$\beta_{\text{AB}} = \frac{\beta_{\text{AA}} + \beta_{\text{BB}}}{2} \tag{5.12}$$

To be consistent with Equations (15.0) and (5.11), β_{AA} is the single-bond energy for element A, and similarly for β_{BB}. Although assumptions are necessary to obtain these energies, Bratsch has compiled a list of the required single-bond energies for the elements.[21] One change was made – the experimental data for N, O and F were corrected for the lone-pair bond weakening effect. The corrections were assigned on the basis of estimates given by Politzer.[22] (The correction is not unambiguous for O and N, because O–O single-bond energies vary from 51 kcal/mol for H_2O_2 to 30 kcal/mol for $(CH_3CO)_2O_2$, with similar variations for N–N. Average values were taken.) The corrected values, with Bratsch's other results, are listed in Table 5.2, for convenience. A few new values have been added.

The β_{AB} values calculated from Table 5.2 are for CN4. Since β is a function of distance, that for CN6 will be less. Looking at the difference in R for CN4 and 6, and using Equation (5.10), it appears that $\beta_{\text{oct}} \approx 0.80\,\beta_{\text{tet}}$. This correction has been used in calculating ΔE_{cov} for CN6.

We must also consider two factors that relate to the ionic model. One is usually called the penetration error. The potential energy in Equation (5.3) is calculated by using point charges for both ions. A more realistic picture has the

Table 5.2 Single-Bond Energies for the Elements[a]

Element	$-\beta$ [kcal/mol]	Element	$-\beta$ [kcal/mol]
Li	25	As	43
Be	55	Se	48
B	73	Br	46
C	85	Rb	12
O	85[b]	Sr	23
N	85[b]	Y	52
F	90[b]	Ag	38
Na	17	Cd	14
Mg	29	In	31
Al	43	Sn	36
Si	55	Sb	34
P	48	Te	38
S	64	I	36
Cl	58	Cs	11
K	14	Ba	19
Ca	24	La	53
Sc	48	Au	52
Cu	46	Hg	10
Zn	17	Tl	23[b]
Ga	34	Pb	28
Ge	45	Mn	39[b]

[a] From Reference 21.
[b] Estimated.

cation as a point charge, but sitting in the electron cloud of the anion. The potential energy is made smaller by this change. Fortunately it is easy to calculate the error with the Slater orbitals used to construct Table 5.1.

The second factor is the radius-ratio effect. Only nearest-neighbor interactions were included in the repulsion part of Equation (5.3). But there is one situation where next-nearest neighbors must be considered: this is the case where one ion, usually the anion, is so much bigger than the other that they are in contact, or strongly overlapping. The values of R_B in Table 5.1 and values of R_0 may be used to assess this possibility.

Table 5.3 shows the experimental values of the cohesive energies and the semi-theoretical values, calculated for pure ionic and pure covalent bonding. Table 5.3 includes most of the available data on AB solids. The omitted examples, mostly alkali halides and alkaline-earth chalcogenides, show no unexpected features.

The ionic energies in Table 5.3 have not been corrected for the penetration error. Table 5.4 gives some sample values of this error. It is negligible for 1–7 salts, and small for 2–6 compounds, except for BeO and BeS. All of the 3–5 cases

Table 5.3 Cohesive Energies of AB Solids

AB	R_0[a]	ΔE_{exp}[b] [kcal/mol]	ΔE_{cov} [kcal/mol]	ΔE_{ion} [kcal/mol]	CN
LiF	2.01	204	114	208	6
BeO	1.65	283	242	277	4
BN	1.56[c]	309	306	375[c]	4
CC	1.54	341	340	–	4
NaCl	2.81	153	81	147	6
MgS	2.60	185	128	160	6
AlP	2.36	193	176	85	4
SiSi	2.35	219	220	–	4
KBr	3.30	142	63	136	6
CaSe	2.90	185	100	172	6
ScAs	2.74	227	139	211	6
CuBr	2.46	133	122	102	4
ZnSe	2.45	125	112	57	4
GaAs	2.45	156	148	−65	4
GeGe	2.44	180	180	–	4
RbI	3.50	125	52	117	6
SrTe	3.17	≈165[d]	83	169	6
YSb	3.08	≈240[d]	134	205	6
AgI	2.81	109	98	85	4
CdTe	2.81	96	98	21	4
InSb	2.81	128	124	−133	4
SnSn	2.81	145	144	–	4
LiI	3.00	128	63	116	6
NaF	2.32	182	108	179	6
CsF	3.01	169	99	164	6
CsI	3.83	127	48	120	8
CuCl	2.35	143	138	114	4
CuI	2.62	123	108	98	4
AgF	2.46	136	134	121	6
AgCl	2.78	127	102	103	6
AgBr	2.89	119	89	90	6
BeS	2.10	200	208	159	4
MgO	2.11	239	157	223	6
MgTe	3.11	132	118	96	4
CaO	2.58	254	153	247	6
CaS	2.85	222	122	193	6
BaO	2.76	235	144	232	6
BaSe	3.30	186	91	185	6
ZnO	1.98	174	177	149	4
ZnS	2.34	147	142	86	4
CdO	2.35	148	138	100	6
CdS	2.52	132	135	71	4

Table 5.3 (*continued*)

AB	R_0[a]	ΔE_{exp}[b] [kcal/mol]	ΔE_{cov} [kcal/mol]	ΔE_{ion} [kcal/mol]	CN
HgS	2.53	95	128	−92	4
MnO	2.22	219	172	193	6
MnS	2.61	185	142	157	6
MnS	2.43	185	178	140	4
BP	1.97	229	234	−12	4
AlN	1.90	267	248	−67	4
AlAs	2.45	178	173	54	4
GaN	1.95	206	226	155	4
GaP	2.36	163	189	−40	4
GaSb	2.64	139	130	−140	4
YAr	2.89	250	146	230	6
InP	2.54	155	152	−27	4
InAs	2.62	145	143	−53	4
CeN	2.50	293	214	296	6
CeP	2.95	≈260[e]	156	230	6
CeAs	3.03	247	146	224	6
LaN	2.65	289	214	285	6
SiC	1.88	295	280	–	4

[a] Data from Reference 23.
[b] Data from Reference 24.
[c] Cubic form; see Table 5.4 for penetration error.
[d] Extrapolated.
[e] Interpolated.

Table 5.4 Representative Penetration Errors[a]

AB	Error [kcal/mol]	AB	Error [kcal/mol]
LiF	0.0	ZnSe	1.5
LiCl	0.06	BN	225
LiBr	0.01	BP	181
AgCl	0.05	AlN	74
BeO	6.2	CeP	8.8
BeS	11.0	AlP	57
MgO	0.05	LaN	5.5
MgS	1.3	GaAs	15.4
CaSe	0.26	ScAs	7.5

[a] Difference between Madelung energies calculated from point charges for anions and from charge clouds for anions.

have appreciable errors, especially those with CN4. The compounds BN and BP, with the diamond structure, have very large errors. For BN the correction for $\Delta E_{\text{ion}} = 375\,\text{kcal/mol}$ reduces the value to $\Delta E_{\text{ion}} = 150\,\text{kcal/mol}$.

The distance between next-nearest neighbors (R_{BB} or R_{AA}) is readily found. It is $1.414R_0$ for CN6, and $1.633R_0$ for CN4. This number may be compared with $2R_B$ or $2R_A$, the distance between two ions when they are in contact. R_B is the anion radius found by X-ray diffraction and listed in Table 5.1. If $2R_B$ is greater than R_{BB}, then the ions are overlapping and large repulsion energies result. It is found that $2R_B$ is comfortably greater than $2R_B$ for all cases in Table 5.3 except for BN, BP, BeS and BeO. These all have CN4 in agreement with the radius-ratio rule. But also, except for BeO, the covalent cohesive energy is much greater than the ionic energy.

Two important conclusions can now be drawn from the data in Table 5.3. The first is that, with few exceptions, compounds with CN4 have ΔE_{cov} much larger than ΔE_{ion}. The reverse is true for CN6. This agrees with expectations: covalent bonding is favored by CN4, and ionic bonding by CN6. The second conclusion is that, with few exceptions, the larger of the two theoretical values, ΔE_{cov} or ΔE_{ion}, agrees reasonably well with the actual cohesive energy, ΔE_{exp}. This is particularly true for CN4, where the agreement is remarkable considering the simplicity of the bonding model. Note that the 4–4 examples are the elements C–C, Si–Si and Sn–Sn, as well as Si–C.

The bonding ability of the ionic model can be improved by varying the net charges on the ions to give the minimum energy.[6] It is possible to write the MO energy equations in terms of the charge x, which can vary from 0 to n for A^nB^{8-n}. When this is done, it is found that $x = 0$ is always a local minimum in the energy, and that $x \simeq 0.9n$ is another local minimum. That is, mixing of the ionic state does not improve the energy of the purely covalent state, but a little mixing improves the energy of the purely ionic state.

A ΔE_{mix} of 5–30 kcal/mol is calculated. It is small for compounds like NaCl and CaO, and large for cases such as CuCl, BeO or MnS. With these corrections, the ionic model gives very good agreement with ΔE_{exp} for all CN6 compounds, except CdO. BeO is predicted to be 95 percent ionic. It has CN4 because of the radius-ratio effect. This is also the case for MgTe.

To summarize, the MO bonding model predicts that most AB solids will fall into two classes: very ionic with the rock-salt structure and 100 percent covalent with a tetrahedral structure. The ionicities are a result of the simplified model and should not be taken too literally. Nevertheless, there are differences in bonding between the solid state and small molecules which are real and which the model correctly identifies. In the solid state, compared with a diatomic molecule, interaction of an ion with the rest of the lattice favors a large value of x. At $x \simeq 0$ there is no stabilization from the lattice. Ionic mixing does not help. Also the metallic element, A, contributes only one to three electrons to four orbitals. Immobilizing these electrons on B to give an ionic state causes an exaggerated loss of covalent bonding.

The reason for two different classes of AB componds is also clearly identified by the model. It is a result of quite different charging energies for metals to the left and to the right in the Periodic Table. Thus, $I_1 = 100$ kcal/mol for K and 178 kcal/mol for Cu; $I_1 + I_2 = 415$ kcal/mol for Ca and 631 kcal/mol for Zn; $I_1 + I_2 + I_3 - 1017$ kcal/mol for Sc and 1320 kcal/mol for Ga. ScN, ScP, ScAs and ScSb all have CN6; GaN, GaP, GaAs and GaSb all have CN4. The non-metalic elements have no part in the structure-making decisions, unless the relative ion sizes are such as to enforce CN4.

The first entries in Table 5.3 are arranged so that the AB compounds formed by elements in the same row of the Periodc Table are grouped together. The values of R_0 for all the CN4 compounds from each row are seen to be nearly constant, including the 4–4 case. Sn–Sn, InSb, CdTe and AgI al have $R_0 = 2.81$ Å, for example. This is consistent with pure covalent bonding in all four cases. The change to CN6 is accompanied by increases of R_0 to 3.08, 3.17 and 3.50 Å for YSb, SrTe and RbI respectively. This suggests a sudden change to ionic bonding. The distances increase steadily because of the decrease of the force of attraction as the charge on the ions becomes smaller.

In molecules the various methods of estimating percent ionic character give results ranging smoothly from 0 percent to near 100 percent. This includes the kind of semi-empirical model used above,[8] as well as much more sophisticated calculations.[25] But in solids there appears to be a discontinuity in ionicity, just as there is a discontinuity in structure. It may be more meaningful to classify solids as either highly ionic or highly covalent, rather than as positioned in a continuous scale.

A few AB solids have body-centered cubic structures, with a CN of 8. An example is CsCl, which is calculated to be much more stable in the ionic model. Again ionic bonding seems to favor a high CN, since the Madelung constant increases, being 1.763 for bcc. This theory proves to be short-lived, however, when we consider bonding in the metals. By definition, the bonding here must be covalent, since identical atoms are bonded.

But the metals are most commonly found with close-packed structures, body-centered cubic (bcc), face-centered cubic (fcc) or hexagonal close-packed (hcp). The number of nearest neighbors, or the CN, is high, being 12 for the close-packed structures and 8(+6) for bcc. So covalent bonding also benefits from high CNs in some cases. At first it seems unhelpful for so many atoms to be within bonding distance, because the non-transition metals, at least, do not have enough valence electrons to bind even four neighbors.

If we try to apply our semi-empirical LCAO–MO method to metals, we get poor results. Taking lithium metal as an example, and using two Li atoms as our unit (just as for C, Si and Ge), we have only two electrons to fill up the four possible MOs. The cohesive energy is only β instead of 4β, the same as it is for the diatomic molecule Li_2. This gives $\Delta E_{cov} = 25$ kcal/mol, using Table 5.2. The value of ΔE_{exp} is 77 kcal/mol. A localized bonding model is not useful for the metals. We must exploit the new features that result because a solid is a very large molecule indeed.

BANDS IN SOLIDS[26,27]

An easy way to see the effect of size on quantized energies is to take the well-known problem of a particle in a box. Start with a single atom in a box of atomic dimensions, and then stack a large number of these boxes to give a macroscopic crystal. Taking Li as an example, and assuming a cubic box of side a, the energy levels for the atomic box are given by

$$\varepsilon = \frac{h^2(n_x^2 + n_y^2 + n_z^2)}{8\,ma^2} \tag{5.13}$$

For the ground state, $n_x = n_y = n_z = 1$. The mass m is the mass of the valence-shell electron, which is the particle. The rest of the atom simply creates a background potential, U_0, constant in the box.

The energy of Equation (5.13) is kinetic energy, which is positive. The potential energy U_0 is negative and can lead to a bound state. To apply Equation (5.13) it is convenient to use atomic units: $h = 2\pi$, $m = e = a_0 = 1$, and 1 a.u. of energy = 27.21 eV. The ionization potential of a Li atom is 5.39 eV. Equating this to the kinetic energy gives an a value of $8.65a_0$ in (Equation (5.13). This is a reasonable result. The average distance of an electron in a 2s orbital of hydrogen is $6.0a_0$, for comparison.

Next we take Avogadro's number (N_0) of Li-atom boxes and put them together to form a cubic crystal of lithium metal. From the density we calculate the side of the cube to be 2.35 cm. Dividing by N_0, we find that each atom is in a box which has a side $a = 5.26a_0$. If each electron was confined to a single atom, the energy would *increase* to 14.56 eV, using Equation (5.13).

Since we need the energy to decrease, showing a cohesive energy, we must allow the valence electrons to move throughout the box. They now move in a constant potential due to the nuclei, the inner-shell electrons and the other valence-shell electrons. This is the free-electron model for metals, due to Sommerfeld.[28]

By imposing a periodic boundary condition in the form

$$\psi(x + L, y + L, z + L) = \psi(x, y, z) \tag{5.14}$$

where L is the side of the crystal, we find that the solutions of Schrödinger's equation are plane waves

$$\psi = (L)^{-3/2} e^{i(k_x x + k_y y + k_z z)} \tag{5.15}$$

The ks are called the wave vector components and are given by $k_x = (2\pi n_x / L)$, and so on.

The one-electron energy levels are

$$\varepsilon = \frac{h^2 k^2}{8\pi^2 m} = \frac{h^2}{8\pi^2 m}(k_x^2 + k_y^2 + k_z^2) \tag{5.16}$$

We must add N_0 electrons to our crystal, in accordance with the Pauli Exclusion Principle. Each level of Equation (5.16) can hold two electrons of opposite spin. The volume element $(2\pi/a)^3$ defines a primitive unit cell in k space. each cell contains one energy level. The ground state will fill all levels from $k = 0$ to a limiting value, k_F. The N_0 electrons will need $N_0/2$ unit cells, or the number lying in a sphere of radius k_F.

The highest energy is that at the surface of the sphere, the Fermi energy, ε_F.

$$\varepsilon_F = \frac{h^2 k_F^2}{8\pi^2 m} = \frac{h^2}{8\pi^2 m}\left(\frac{3\pi^2 N_0}{V_0}\right)^{2/3}$$

The molar volume, V_0, is simply $N_0 a^3$. Also, the average energy per electron can be shown to be equal to 3/5 of ε_F, so we can write

$$\bar{\varepsilon} = \frac{3h^2(3/\pi)^{2/3}}{40ma^2} \tag{5.18}$$

For lithium metal this is 2.82 eV, which is to be compared with the average energy of the separate Li atoms of 5.39 eV.

There has been a reduction in energy of 2.57 eV, or 59.3 kcal/mol. For two atoms this means a cohesive energy of 119 kcal/mol, which is to be compared with $\Delta E_{exp} = 77$ kcal/mol. Of course we must also consider the change in U_0 to obtain a more complete story, but the essential point has been established. Putting a large number of electrons in a large box produces a reduction in energy, or a delocalization energy.

Whereas physicists are comfortable with plane waves, chemists will usually prefer an explanation based on LCAO theory. This is the tight-binding approach, in solid-state terminology. The result may be anticipated by our general knowledge of valence theory. Figure 5.1 shows the building up of energy levels as we increase the number of atoms bonded together. For one valence orbital per atom, one more MO is formed for each atom added. For a linear chain, the lowest-energy MO is pushed down and the highest MO is pushed up with each addition of an atom. Also the levels are more closely packed, so that for an infinite number of atoms a continuous band of levels is formed.

If there is one electon per valence orbital, then only the bottom half of the band will be filled, since we can put two electrons of opposite spin into each level. Thus all the bonding orbitals will be filled and all the anti-bonding orbitals will be empty. The result is a net stabilization of the system, or a cohesive energy. If there are several valence orbitals on each atom with different energies, then several bands will be formed, which may or may not overlap.

This picture can be made more quantitative, at least at the level of simple Hückel MO theory. Take a linear chain of H atoms, or a linear chain of CH units, with a constant spacing, a, between the units. The energy of the levels in such a case is given by

$$\varepsilon_j = \alpha + 2\beta\cos(2\pi j/(N+1)) \tag{5.19}$$

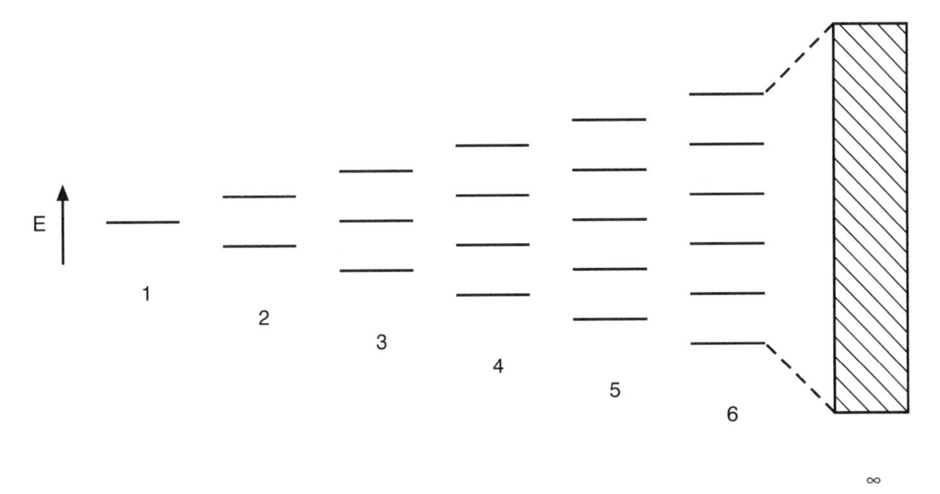

Figure 5.1 The building up of an energy band by repeated addition of atoms

where the index j refers to the MO, or level, and N is the number of atoms. defining a wave vector, $k = 2\pi j/Na$, the energies are given by

$$\varepsilon_k = \alpha + 2\beta \cos ka \tag{5.20}$$

when N is large.

The wave vector forms a reciprocal space as before and has continuous values in the range $-\pi/a < k < \pi/a$. These limits define the (first) Brillouin zone of a crystal. The point $k = 0$ is the zone center and $k = \pm\pi/a$ are the zone edges. The variation of the energy with k is called the dispersion of the band. Figure 5.2 shows the dispersion given by Equation (5.20). Only the range from 0 to π/a need be shown. The range from 0 to $-\pi/a$ is just the mirror image. Figure 5.2 also shows the Fermi level, $k_F = \pi/2a$, which is the highest filled level if the levels are doubly occupied.

The energy per electron ranges from $\varepsilon_0 = \alpha + 2\beta$ to $\varepsilon_F = \alpha$. We can find the average energy by simply counting the energies of the occupied levels:

$$\bar{\varepsilon} = \frac{2a}{\pi}\int_0^{k_F} \varepsilon(k)\mathrm{d}k = \frac{4\beta}{\pi} + \alpha \tag{5.21}$$

Since the levels are nearly continuous, this can be done by integration. This energy should be compared with the energy of an electron in the two-atom case, $\bar{\varepsilon} = \alpha + \beta$. There is a delocalization energy of $0.27\,\beta$/electron. Note that in Hückel theory, there is no repulsion energy term. The Hückel β is one-half of the β values in Table 5.2.

This analysis was for a one-dimensional solid. The extension to three dimensions is not so easy for real structures, but can be done for a simplified model of a metal. Imagine the case of a simple cubic lattice such as is shown in Figure 5.3.

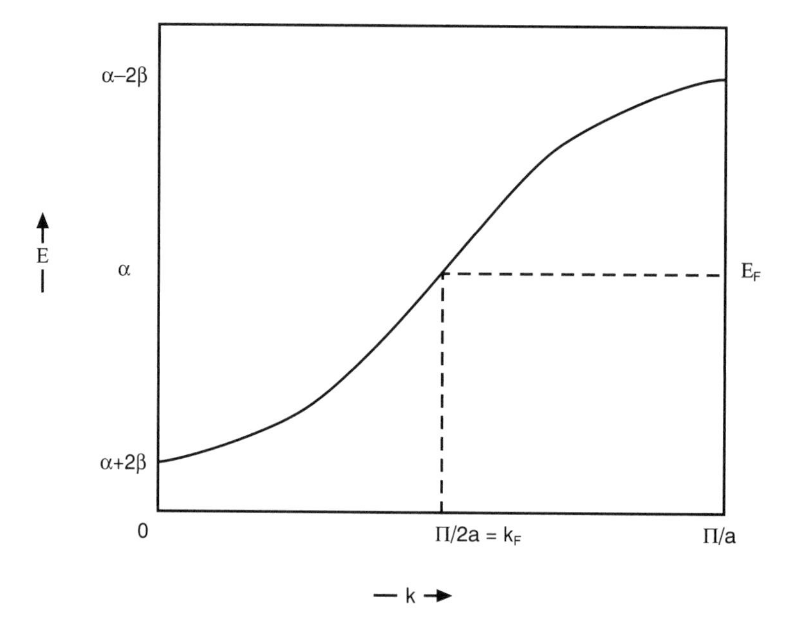

Figure 5.2 Dispersion of the one-dimensional energy band formed by overlap of the s or pπ orbitals of an infinite chain of atoms

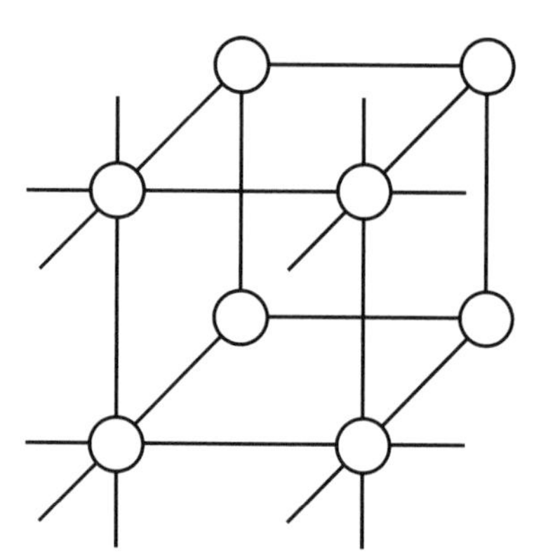

Figure 5.3 A simple cubic structure for a solid which can form chains in the x, y and z directions

The coordination number is only 6, but if the valence orbital is an s orbital, then we have three infinite chains for each atom, running in the x, y and z directions. The energies are[29]

$$\varepsilon(k) = \alpha + 2\beta(\cos k_x a + \cos k_y a + \cos k_z a) \tag{5.22}$$

Now the average energy becomes

$$\bar{\varepsilon} = \alpha + \frac{12\beta}{\pi} = \alpha + 3.82\beta \tag{5.23}$$

For two lithium atoms, the cohesive energy is 7.64β, compared with 2β for an Li_2 molecule. This can account for the difference found experimentally: 25 kcal/mol compared with 77 kcal/mol. One should also consider internuclear separations, 2.67 Å in the diatomic molecule and 3.03 Å in the metal, so that β in the metal is less than in the molecule.

The dispersion curve for the $\varepsilon(k)$ of Equation (5.22) is more complicated than Figure 5.2, since k_x, k_y and k_z can have different values. The convention is to show the energies at various selected points in the Brillouin zone. These are labeled as Γ, M, K and X, and are called symmetry points.

Returning to the one-dimensonal case again, the wave functions may be written as

$$\phi(k) = (N)^{-1/2} \sum_{p=1}^{N} \exp(ikR_{\mathrm{p}})\psi(r - R_{\mathrm{p}}) \qquad (5.24)$$

$\boldsymbol{R}_{\mathrm{p}}$ is the vector which translates a point in one unit cell to the corresponding point in another. The index p identifies the unit cell, or the atom. ψ is an atomic orbital, which is the same for all atoms. The $\phi(k)$s are called Bloch functions and are simply symmetry-adapted linear combinations of the orbitals ψ.

In three dimensions the exponential in Equation (5.24) must be written as a dot product, $\exp(i\boldsymbol{k}\cdot\boldsymbol{R}_{\mathrm{p}})$. Also, in real solids the atoms of the unit cell usually contribute more than one valence orbital. Then one has a set of Bloch functions and the crystal orbitals are linear combinations of the Bloch functions. Also, the atoms of the unit cell are usually not the same. At least in the case of the AB solids discussed in the previous section, we can show how this affects the results.

Take the one-dimensional case as a start. There is an infinite chain of alternating A and B atoms, with a constant spacing a. The unit cell must now contain two atoms. The energy levels are given by

$$\varepsilon(\boldsymbol{k}) = \frac{\alpha_{\mathrm{B}} + \alpha_{\mathrm{A}}}{2} \pm \frac{1}{2}\sqrt{(\alpha_{\mathrm{B}} - \alpha_{\mathrm{A}})^2 + 16\beta^2 \cos ka} \qquad (5.15)$$

For any value of $\boldsymbol{k}$ there are two roots, one bonding and one anti-bonding. Figure 5.4(a) shows the two energies plotted as a function of k, for typical values of α_{B}, α_{A} and β.

When $\alpha_{\mathrm{B}} \gg \alpha_{\mathrm{A}}$, Equation (5.25) may be expanded as[29]

$$\varepsilon_1 = \alpha_{\mathrm{B}} + \frac{4\beta^2 \cos ka}{(\alpha_{\mathrm{B}} - \alpha_{\mathrm{A}})} \qquad (5.26)$$

$$\varepsilon_2 = \alpha_{\mathrm{A}} - \frac{4\beta^2 \cos ka}{(\alpha_{\mathrm{B}} - \alpha_{\mathrm{A}})} \qquad (5.27)$$

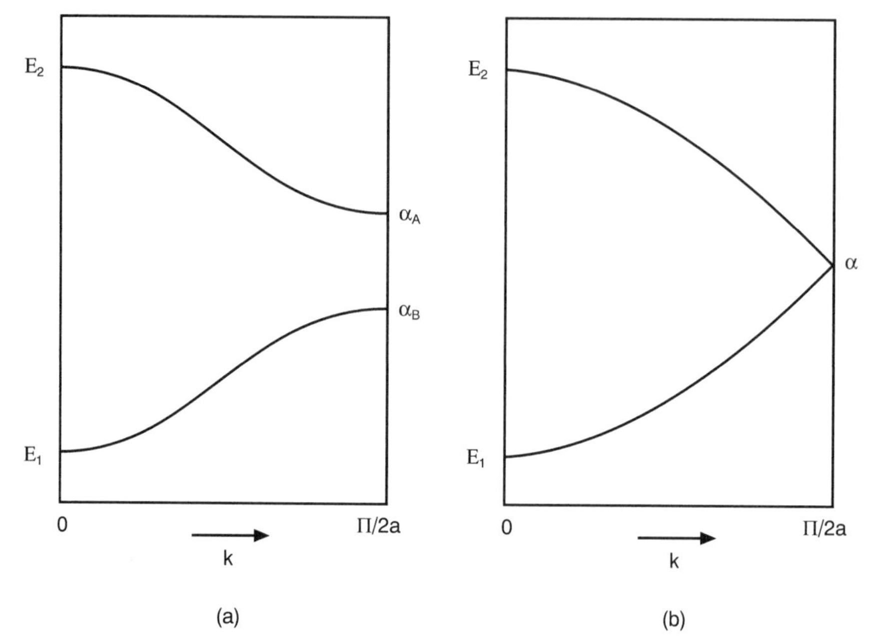

Figure 5.4 (a) Dispersion of the bonding (lower curve) and anti-bonding (upper curve) levels of an AB solid; (b) the same when $\alpha_A = \alpha_B$

Important new features are that there are now two bands, and that a band gap exists at $k = \pi/2a$. The gap size is $|\alpha_B - \alpha_A|$. Of course when A = B, as in C–C or Si–Si, the gap must disappear, as shown in Figure 5.4(b). Although this looks different from Figure 5.2, it is the same actually, as may be seen if one folds Figure 5.2 back on itself in the middle. The difference exists because there are now two atoms in the unit cell instead of one.

We also can see that, if β is very small, the gap would be constant, independently of k. But it is a dependence of ε on k that produces the band width, or dispersion. Compounds which are highly ionic will have narrow bands. The ground state will have the electrons concentrated on B, the non-metallic atom. Excited states result from the transfer of an electron from B^- to A^+, forming A,B.

In three dimensions there will be two sets of four overlapping bands for the compounds in Table 5.2. In the rock-salt structure linear chains exist for the p_x, p_y and p_z orbitals and Equation (5.26) and (5.27) are appropriate. The s orbital would give somewhat more delocalization energy, as in Equation (5.22). All four of the stable bands will be filled completely by the eight valence electrons of the AB unit. These are called the valence bands. All four of the unstable bands are completely empty. They are called the conduction bands, for historical reasons.

In the limit where $\beta = 0$, we have the completely ionic case. The average energies of the filled bands are given by Equations (5.3) and (5.4). In terms of the model leading to Equation (5.26), the energy per electron is equal to α_B, which then must include the terms in Equations (5.4) and (5.6), assuming that a repulsive potential has been added. The average energy of the empty, or conduction, band will be equal to α_A, as in Equation (5.27). This must also include terms in I_n and A_n, as well as coulombic energies appropriate for an A,B unit in a matrix of A^+, B^-.

In the wurtzite or zinc blende structure, we do not have linear chains of nearest neighbors. Usually there is strong mixing of the s and p bands. The actual calculation of the band structure of, say, diamond, is very complicated,[30] but a simple model gives the essential features. Refer back to Equation (5.9), showing the energy lowering for eight electrons in four bonding orbitals as equal to $|8\beta|$. Not shown is the fact that there are also four anti-bonding orbitals, with an energy increase of $-\beta$ per electron.

Just as we argued earlier that we could take linear combinations of the crystal orbitals to form localized orbitals, we can now reverse the argument and take linear combinations of the localized orbitals to form the crystal orbitals. If we only combine orbitals which are completely filled, or completely empty, we will not change the average energy. Thus in diamond we will generate a band of filled orbitals with an average energy of $(\alpha + \beta)$, and a band of empty orbitals with an average energy of $(\alpha - \beta)$ per electron.

Estimating the width of the bands is more difficult. However, we can obtain some idea by looking again at Figure 5.2. Treating the filled and empty levels as the valence band and the conduction band, we see that the average energy difference in $8\beta/\pi$, or 2.5β. The width of each band is 2β. Therefore, for covalent bonding, the band width is of the same order as the *average* energy gap.

The conclusion is that band widths will be appreciable, but usually not great enough to close the gap between the full and empty bands. Figure 5.5 shows schematically the three situations that finally emerge. They are classified according to their electrical resistance. Only the highest filled band and the lowest empty band are shown. The vertical scale is energy, but the horizontal scale has no meaning.

Conductors, such as the metals, are characterized by a partially filled band, so that the highest filled level and the lowest empty level are essentially at the same energy, the Fermi energy. Insulators have a large residual gap between the valence and conduction bands. Examples are ionic compounds, but also some covalent compounds such as diamond. Semiconductors have a small gap between the bands. Most of the covalent compounds in Table 5.3 fall into this class.

The electrical conductivity of electrons in a solid depends on the ability of an electron to move to a higher energy level when accelerated by an electric field. The energy change is very small, so that only partially filled bands can conduct. In semiconductors thermal energy will promote a few valence-band electrons into the conduction band. These electrons can now move in the field. So can the electrons in the valence band whose energies are just below the levels of the promoted electrons.

Figure 5.5 enables us to find the quantities μ and η, as defined in Equation (5.1), and as approximated by I and A. In the solid the ionization potential is

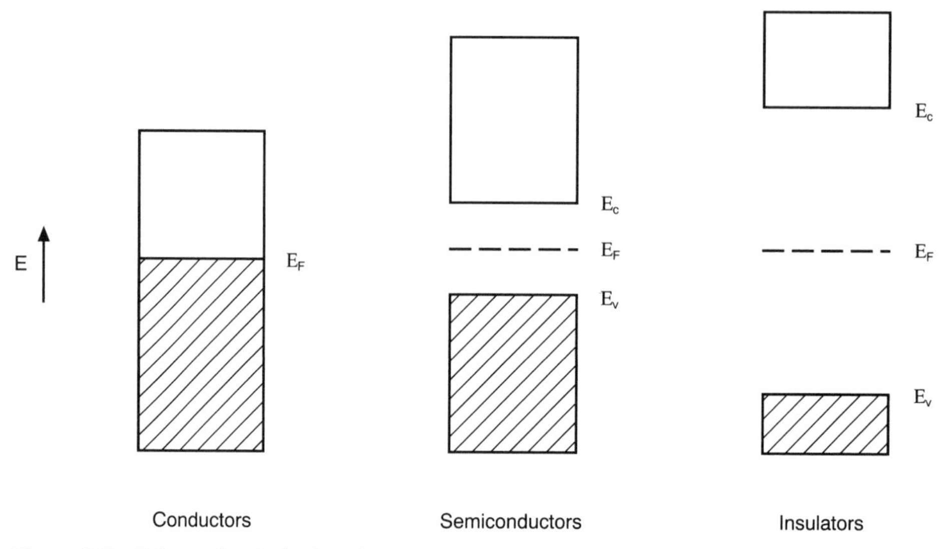

Figure 5.5 Schematic of the band structures for conductors (metals), semiconductors and insulators. The shaded areas show the occupied levels. The lower band is the valence band, and the upper is the conduction band

called the photoelectric threshold and is experimentally measurable. It will be equal to $-\varepsilon_V$, the highest filled level in the valence band. The electron affinity, A, is not measurable as a rule, but is equal to $-\varepsilon_c$, the lowest empty level in the conduction band. We have

$$\mu = \frac{(\varepsilon_V + \varepsilon_c)}{2}; \qquad \eta = \frac{(\varepsilon_c - \varepsilon_V)}{2} = \frac{E_g^\circ}{2} \tag{5.28}$$

at the orbital level of approximation.

Thus twice the chemical hardness is equal to the minimum energy gap, E_g°, an important property in solid-state physics. In the case of metals, $\varepsilon_V = \varepsilon_c = \varepsilon_F$. Therefore $\mu = \varepsilon_F$ and $\eta = 0$. Actually the hardness is not exactly zero, since there is a small energy difference between successive levels. The work function, Φ, is defined as the minimum energy required to remove an electron from the Fermi level. For metals we have $-\Phi = \varepsilon_F = \mu$.

For insulaors and semiconductors the Fermi energy is also set equal to μ, and therefore is in the forbidden region between the bands, as shown in Figure 5.5. No electron is actually at the Fermi level. The reasoning for setting $\varepsilon_F = \mu$ follows from the Fermi distribution law. This gives the probability of occupation of an electron level as a function of temperature:[31]

$$f(\varepsilon) = \left(\exp\left(\frac{\varepsilon - \mu}{kT}\right) + 1\right)^{-1} \tag{5.29}$$

where μ is the ordinary chemical potential of the electrons, as well as our electronic chemical potential. If $\varepsilon = \varepsilon_F = \mu$, then the probability is equal to one-half, which is correct for a level at the occupation boundary.

Figure 5.6 shows $f(\varepsilon)$ plotted against ε at absolute zero and at a finite temperature. At the higher temperature a few electrons are promoted above μ and a few holes are created below μ, as already mentioned. At absolute zero, $f(\varepsilon)$ is a step function equal to 1 for $\varepsilon < \mu$, and zero for $\varepsilon > \mu$.

In electrochemistry, in the solid state, there are both μ, the ordinary chemical potential of the electrons, and μ', the electrochemical potential.[32] The two are related by

$$\mu' = \mu - \phi \tag{5.30}$$

where ϕ is the electrostatic potential in the bulk solid. The work function is given by[33]

$$\Phi = -\mu + (\phi - \phi') \tag{5.31}$$

The added term is the potential difference between a point just inside the surface and one just outside the surface of the solid.

The term $(\phi - \phi')$ is also the potential due to the surface dipole layer. It is a function of the condition of the surface and is not a constant. The problem in elctrochemistry is that the emphasis is on measuring electrical potentials. In addition to the potential due to different values of μ for two electrodes, the

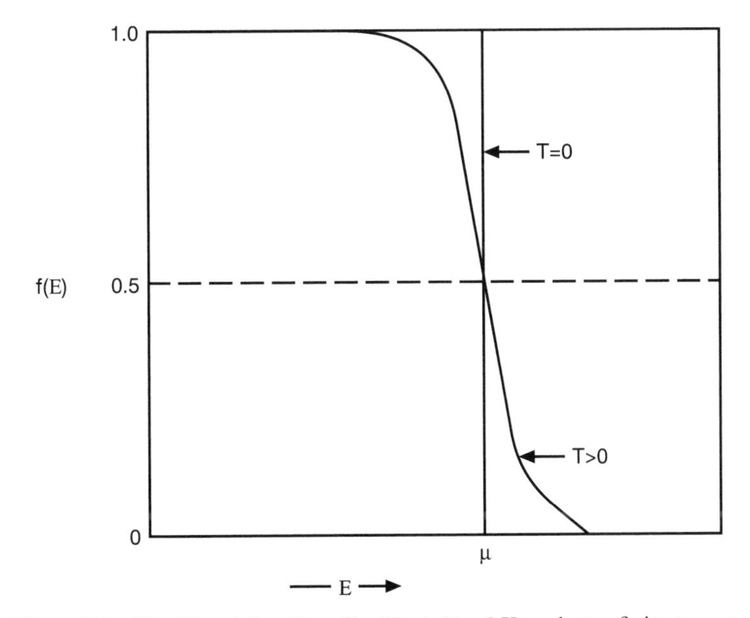

Figure 5.6 The Fermi function $f(\varepsilon, T)$ at $T = 0\,\mathrm{K}$ and at a finite temperature

measured voltages will include surface potentials, which are often substantial. However, in terms of energies they are negligible and are usually ignored in solid-state physics. In the same way, a slight charging of an electrode can produce a measurable change in voltage, but will have little effect on the total energy, or the average energy of an electron in the bulk.

When two solids are put in contact with each other, there will be a contact potential, called the Volta potential, V. It is given by

$$V = (\Phi_1 - \Phi_2) \tag{5.32}$$

the difference in work functions. The potential is due to the transfer of electrons between the two solids. The condition of equilibrium is that the electrochemical potentials be equal, i.e.,

$$\mu_1' = \mu_2' = \mu_1 - \phi_1 = \mu_2 - \phi_2 \tag{5.33}$$

The best-known property of a solid electrode is its redox potential on the hydrogen, or relative, scale. Putting the potential on an absolute scale, e.g.,

$$\text{Ag(s)} = \text{Ag}^+\text{(aq)} + \text{e}^-\text{(g)} \qquad I' \tag{5.34}$$

it is obvious that there is no simple relationship between I' and any of the quantities discussed above. Reaction (5.34) does correspond to a definite change in energy, or free energy; at standard temperature. Attempts to measure it, or calculate it, will always run into problems, such as the existence of surface potentials.[33]

INSULATORS AND SEMICONDUCTORS

The purpose of this section is to add the DFT-based concepts of μ and η, as defined in equation (5.28), to the existing treatment of solid-state insulators and semiconductors. Also we will use the very simple theory of bonding developed earlier for ionic and covalent bonding to predict, or rationalize, certain properties of importance in solid-state physics.

Of course, we should also mention some of the many other uses of DFT in the study of solids. Particularly important are the detailed *ab-initio* calculations based on DFT.[34] These calculations have led to values of the cohesive energies, bulk moduli, energy bands, densities of states and magnetic properties. Generally speaking, the results have been successful, though oftentimes there are quite large errors (~20%) in the cohesive energies. Metals, insulators and semiconductors have been considered.[25] For the latter two cases there is difficulty in calculating the band gaps, which are always too small.[35] This is attributed to errors in the conduction-band energy levels, which are mainly unoccupied and similar to the LUMOs of molecules.

Returning to a consideration of μ and η, Equation (5.28) came from the assumption by Parr that the chemical potential of an electron in a molecule was of the same nature as the chemical potential of an electron in a solid. This was accompanied by a finite difference approximation to a derivative and the use of Koopmans' theory, relating orbital energies to I and A.

In view of these uncertainties, it is fortunate that an independent proof of Equation (5.28) exists in solid-state physics.[36] Equation (5.29), the Fermi distribution function, is used to count the number of electrons in the valence band and in the conduction band. The result is that

$$\mu = \frac{\varepsilon_c + \varepsilon_v}{2} + \frac{kT}{2} \ln\left(\frac{N_v}{N_c}\right) \tag{5.35}$$

where N_v and N_c are called the effective number of states per unit volume in the valence and conduction bands, respectively. Their ratio is not very large, and since kT is only 0.02 eV at room temperature, the second term in equation (5.35) can usually be neglected.

Actually, a similar slight dependence of μ on the temperature is found for molecules.[37] It involves the ratio of the degeneracies of M^+ and M^-. Looking at equation (5.29) again, we see that if the total number of electrons is to remain constant, then μ must change as the temperature changes. This shows up in Equation (5.35). In the case of semiconductors and insulators, the convention is used that $\varepsilon_F = \mu$, at all temperatures. For metals, the convention is that the Fermi energy is always the value at 0 K.

The chemical potential or its equivalent, the work function, has long been used to measure the electron-donating and -accepting power of a solid. For two solids in contact, the difference determines the flow of electrons. This is an important property for semiconductor devices.[38] The interaction between a solid and a molecule also is influenced by the differences in their electronic chemical potentials. However, such interactions almost always occur at a surface, which brings in important new effects. A brief discussion of surfaces will be given later.

It seems natural to consider E_g° (i.e., twice the chemical hardness), if transfer of electrons is the critical property. Certainly the fact that the gap is zero for metals accords with the great electrical conductivity of these solids. Also, the size of the gap determines the conductivity of semiconductors and insulators. In general, the conductivity, C, is given by

$$C = nm_e + pm_h \tag{5.36}$$

where n is the concentration of conduction-band electrons and p is the concentration of valence-band holes. In an intrinsic semiconductor, or insulator, $n = p$.

The quantities m_e and m_h are the mobilities of the electrons and the holes. They are proportional to the time between scattering events and inversely proportional to the so-called effective masses. The holes move in an electric field by neighboring electrons jumping into the hole, and creating a new hole. As the temperature increases, the mobilities decrease, but n and p increase. At room temperature the increase in n and p for metals is less than the decrease in m_e and m_h, and C decreases with increasing temperature.

For insulators and semiconductors the situation is reversed, and the conductivity increases with temperature. The concentrations of the charge carriers are given by

$$n = N_c \exp\left(\frac{\mu - \varepsilon_c}{kT}\right); \qquad p = N_v \exp\left(\frac{\varepsilon_v - \mu}{kT}\right) \tag{5.37}$$

If this is combined with $n = p$ and Equation (5.35), we obtain

$$n = p = (N_c N_v)^{1/2} \exp\left(\frac{-E_g^\circ}{2kT}\right) \tag{5.38}$$

The value of the energy gap may be found by plotting the log of C against $1/T$.

The minimum energy gap is also the important factor for other properties of a solid which depend on the electrons in the conduction band. These include the Pauli spin paramagnetism, and the (small) contribution of the electrons to thermal conductivity. All of these properties are due to extremely small concentrations of free electrons. Thus for silicon, where $E_g^\circ = 1.1$ eV, the number of conduction electrons is only $2 \times 10^{10}/\text{cm}^3$, compared with an atom concentration of $5 \times 10^{22}/\text{cm}^3$. This is for a sample where impurity concentrations have been reduced to 1 part in 10^{12} by zone refining.

But there are many other properties of a solid which are little affected by such small numbers. These are the properties where the response to a change in conditions requires a change in *all* of the valence electrons, not just the few near the Fermi level. Certainly this would be the case for the cohesive energy. The appropriate energy gap in such cases would be that between the *average* energy in the valence band and the *average* energy in the conduction band, so in place of Equation (5.28) we would have

$$\bar{\mu} = \frac{\bar{\varepsilon}_v + \bar{\varepsilon}_c}{2}; \qquad \bar{\eta} = \frac{\bar{\varepsilon}_c - \bar{\varepsilon}_v}{2} = \frac{\bar{E}_g}{2} \tag{5.39}$$

Table 5.5 shows experimental values of $\bar{E}_g$, E_g° and the cohesive energies (from Table 5.3) for a number of AB compounds. The average energy gap results are those calculated from experimental data on high-frequency dielectric constants for the crystals. Later we will compare these values of $\bar{E}_g$ with those calculated from our earlier bonding models.

Table 5.5 Comparison of Energy Gaps with Cohesive Energies

Solid	$\bar{E}_g$ [eV][a]	ΔE_{coh} [kcal/mol][b]	E_g^0 [eV][c]
C	13.6	341	5.4
SiC	9.2	295	3.1
Si	4.8	219	1.1
Ge	4.3	180	0.7
Sn	3.1	145	0.0
BN	15.3	309	4.6
AlN	11.0	267	3.8
AlP	5.6	193	3.0
AlAs	5.2	178	2.2
GaN	10.7	206	3.3
GaP	5.6	163	2.2
GaAs	5.2	156	1.3
GaSb	4.1	139	0.8
InP	5.2	155	1.3
InAs	4.6	145	0.4
ZnO	11.8	174	3.7
ZnS	7.8	147	3.5
ZnSe	7.1	125	2.6
CdO	9.0	148	2.5
CdS	7.1	132	2.4
CdTe	5.4	96	1.5
MgO	15.8	239	7.3
CaO	15.3	254	6.9
CaS	9.6	222	5.4
CaSe	8.5	185	5.0
SrO	13.9	240	5.3
SrTe	7.1	165	4.0
LiF	24.0	204	13.6
LiCl	12.2	165	9.4
LiBr	10.0	149	7.6
LiI	7.8	128	~6.0
NaF	21.5	182	11.6
NaCl	12.2	153	8.5
KBr	9.5	142	7.4
RbI	7.3	125	6.1
AgCl	5.4	127	3.0
AgI	6.5	109	2.8
CuCl	9.6	143	3.3
CuBr	8.0	133	3.0
CuI	6.6	123	3.1

[a] References 39b and 45.
[b] See Table 5.3.
[c] References 43, 45 and 46.

A reason for selecting the dielectric constant as the experimental basis for $\bar{E}_g$ is that the high-frequency value, ε_∞, is due entirely to the polarization of the electrons. Also ε_∞ is equal to the square of the refractive index and is easily measured. There is a simple relationship between the average energy gap and ε_∞[39]

$$\varepsilon_\infty \simeq 1 + 4\pi N/\bar{E}_g^2 m \tag{5.40}$$

where N is the number of valence electrons per unit volume and m is the electron mass.

Another reason for using ε_∞ as the source of $\bar{E}_g$ is that the second term in Equation (5.40) is the contribution due to the polarization of the valence electrons, i.e., the response of all these electrons to the perturbation of a weak electric field. In perturbation theory this response is inversely proportional to the differences in energy between the various excited states and the ground state. It is common practice to replace the various energy differences by a single average value, our $\bar{E}_g$. An electron at the bottom of the valence band will contribute almost as much of the dielectric constant as an electron at the top. Looking at Equation (5.38), such an electron contributes nothing to the conductivity at room temperature.

An early use of Equation (5.40) was by Phillips and van Vechten.[40] They applied it to derive a scale of percent ionic character. Their assumptions were such that a continuous scale was generated. While such a scale may be questioned for solids, there was also an important conclusion: the structures of the AB compounds depended entirely on the ionicity. Low ionicity led to CN4, and high ionicity to CN6 or 8. This is consistent with our earlier analysis based on cohesive energies.

Our immediate concern is whether $\bar{E}_g$, or possibly E_g°, serves as a suitable measure of chemical hardness just as $(I - A)$ does for molecules. Examination of the data in Table 5.5 shows that it does. There is a good correlation btween $\bar{E}_g$ and the cohesive energy, as long as related solids are compared. That is, the 4–4 compounds show $\bar{E}_g$ falling just as ΔE_{coh} does. The alkali halides also are correlated with each other, but not with the 4–4 cases. In the 2–6 examples, we can compare the CN6 compounds with each other, but not with the CN4 cases, which have their own relationship. The 3–5 solids form their own family for CN4, but there are no $\bar{E}_g$ data for the ionic 3–5 cases, which probably belong to a different family.

Actually, the same correlations are found for the values of E_g° and ΔE_{coh}, but it is not as good, simply because of the smaller range of E_g°. This behavior is not unexpected. Note that $\bar{E}_g$ is always much greater than E_g°, as it must be. As an additional test of the direct relationship between hardness and stability, it has been shown that, when different structures are possible for a solid, the more stable structure has the largest energy gap.[41]

We can also obtain a set of numbers for $\bar{E}_g$ by using the simplified bonding model which led to Table 5.3. To do this, we will simply calculate the change in energy in promoting one electron from the localized bonding orbital to the localized anti-bonding orbital. Before proceeding, it will be helpful to review the various energy levels of a solid to which we have referred, and to which labels have been assigned. Figure 5.7 does this schematically. It serves for both semiconductors and insulators, depending on the size of the gap.

What is usually available from experiment is the ionization potential, I°, which comes from photoelectric or thermionic emission, and the band gap, E°_g. From these data the Fermi energy and the electron affinity, A°, can be calculated. The Fermi energy is equal to the work function, Φ, with change in sign. Additional data can be obtained from the vis-UV absorption spectrum, but this is a more difficult problem of interpretation.

To test our bonding model, as before, separate calculations must be made for the purely ionic case and the purely covalent case. However, from the outset we will assume that CN4 means covalent, and CN6 means ionic. The main test will be to see if we can match the values of $\bar{E}_g$ in Table 5.5 with theoretical results, but we can also try to match I° and A°. The calculation of E°_g is not possible with our bonding model.

We will start with the ionic case and calculate the average values $\bar{I}$ and $\bar{A}$ for NaCl. $\bar{I}$ refers to the conversion of Cl^- to Cl in the lattice. The energy required consists of three parts: the electron affinity of the chlorine atom, the loss of the coulombic energy of one ion, and a small energy lowering due to the polarization of the medium after the electron loss:

$$Cl^-(s) = Cl(s) + e^-(g) \tag{5.41}$$

$$\Delta E = 3.62 + \left(\frac{1.748}{2.81}\right)14.39 - 1.5 = 11.1\,\text{eV} = \bar{I} \tag{5.42}$$

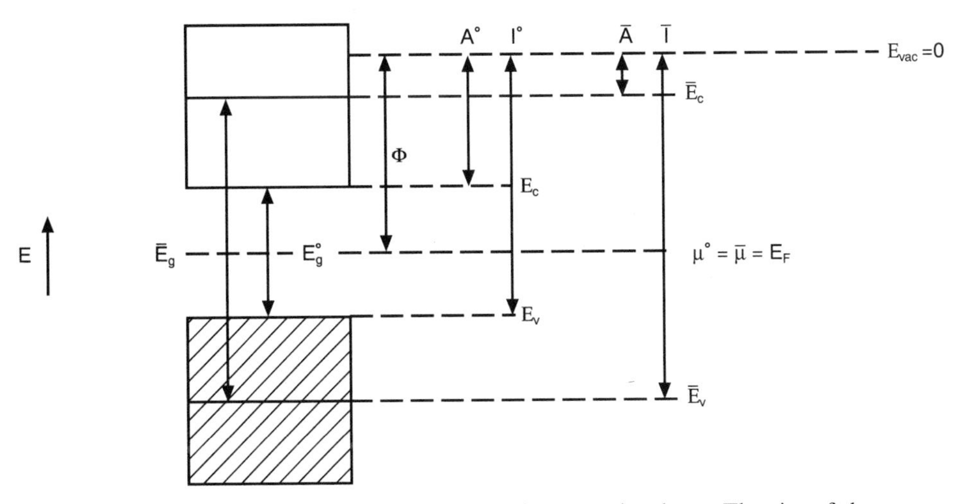

Figure 5.7 Relevant energy levels for a semiconductor or insulator. The size of the arrows gives the magnitude of energy differences

The number 14.39 converts the coulombic energy to electron-volts, if R_0 is in Ångströms.

The polarization energy arises when a positive charge is converted to a neutral entity.[42] This seems contradictory to the usual case where a charge is formed from a neutral species, followed by a polarization of the medium. However, in both cases there is a change in the electric field. In the crystal lattice, when perfect, the fields vanish at each ion by symmetry. If one Cl^- is converted to Cl, there is an unbalanced field left at each of the six adjacent Na^+ and they are polarized, lowering the energy.

The electron affinity also has three parts: an energy lowering due to converting Na^+ to Na, the loss of coulombic energy and a polarization term

$$Na^+(s) + e^-(g) = Na(s) \tag{5.43}$$

$$\Delta E = -5.14 + \left(\frac{1.748}{2.81}\right) 14.39 - 2.5 = 1.3\,\text{eV} = -\bar{A} \tag{5.44}$$

These numbers are to be compared with experimental values of $I^\circ = 9.0\,\text{eV}$ and $A^\circ = +0.5\,\text{eV}$ for NaCl. They are not the same since they differ by one-half the band widths, as Figure 5.7 shows. The valence band width is 4.1 eV for NaCl,[43] which makes $\bar{I}$ and I° agree. The conduction band width needed to make $\bar{A}$ and A° agree is 3.6 eV. Actually this piece of information is not readily available from experiment.

The band width of the valence band is due to exchange interactions between neighboring chloride ions. The conduction band is rather indefinite since its center is near zero energy. Thus many other states, such as excited states of Cl^-, are overlapping; there is a continuum of levels available.

The next step is to combine reactions (5.41) and (5.43) into a single process

$$Na^+(s) + Cl^-(s) = Na(s) + Cl(s) \tag{5.45}$$

By an electron jump, we have formed an electron–hole pair, to use the solid-state terminology. This can occur in two ways: the two atoms can be widely separated, or they can be adjacent. In the first case, the energy required is less, by just the coulombic energy of a neighboring Na^+ and Cl^-, or 5.1 eV. This means that the electron and the (positive) hole attract each other by this amount.

The energy of reaction (5.45) is another way of clculating $\bar{E}_g$, but we have two values: 12.4 eV and 7.3 eV. The latter figure can be discounted, however, since the electron is not free to conduct electricity. Therefore it is not in the conduction band. Such a bound electron–hole pair is called an exciton. We can also measure the energy of reaction (5.45) by vis–UV absorption spectroscopy. For NaCl there is a rather sharp band at 7.8 eV, which is not photoconductive, followed by a broad continuum. Photoconductivity starts at 8.5 eV, but does not become strong until about 11 eV and thereafter, as the absorption increases.

Accordingly, we identify 12.4 eV as the value of $\bar{E}_g$ from our bonding model. This closely agrees with $\bar{E}_g = 12.2\,\text{eV}$, found from the dielectric constant. Table 5.6 contains comparisons of the same kind for ionic compounds. The agreement is surprisingly good, since the excited states contributing to the polarizability need not be the same as the lowest excited states in the UV spectrum.

For small molecules there was a difference between the spectroscopic HOMO–LUMO gap and the value of $(I - A)$ using experimental values of the ionization potential and elecron affinity. The difference was the interelectron repulsion of an electron in the HOMO and one in the LUMO. This is not a factor in the solid state, since the electrons are now in crystal orbitals. These are spread over the entire crystal, and the mean repulsion of two electrons is negligible.

Turning next to the covalent case, silicon is taken as an example. For the 4–4 compounds we have

$$\bar{\varepsilon}_v = \alpha + \beta; \qquad \bar{\varepsilon}_c = \alpha - \beta \tag{5.46}$$

so that $\bar{E}_g = 2\beta = 4.78\,\text{eV}$, with β taken from Table 5.2. This means that we are ignoring any changes in the repulsion energy term, as given in Equation (5.10). This was also done in the previous calculations on ionic compounds.

Finding $\bar{I}$ and $\bar{A}$ separately is rather different, since we cannot assume that $\bar{I} = -\bar{\varepsilon}_v$ and $\bar{A} = -\bar{\varepsilon}_c$. There are two changes: one is that, since ions are actually formed, we must consider the consider the solvation energy of the ions in a medium with a dielectric constant. The Born equation seems a reasonable approximation

$$\Delta E_{\text{solv}} = -\frac{q^2}{2R_0}\left(\frac{\varepsilon_s - 1}{\varepsilon_s}\right) 14.39\,\text{eV} \tag{5.47}$$

where ε_s is the static dielectric constant of silicon. The ionic radius, R_0, is taken to be the same as the interatomic spacing in silicon for both Si^+ and Si^-. This gives $\Delta E_{\text{solv}} = -2.81\,\text{eV}$ for both cases.

Table 5.6 Comparison of $\bar{E}_g$ from Dielectric Constant with Theoretical Spectroscopic Gap

CN6			CN4		
Solid	$\bar{E}_g$ [eV]	$(I - A)_s$ [eV]	Solid	$\bar{E}_g$ [eV]	$(I - A)_s$ [eV]
LiF	24.0	18.3	C	13.6	7.4
LiI	7.8	10.0	Si	4.8	4.8
NaCl	12.2	12.4	Sn	3.1	3.1
KBr	9.5	10.5	BN	15.3	9.8
RbI	7.3	9.8	AlP	6.0	5.6
AgCl	8.4	10.1	GaAs	5.2	5.4
CaO	15.3	11.4	ZnS	7.8	7.2
MgSe	7.2	6.6	CuCl	9.6	9.3

For the ionization potential we find

$$\mathrm{Si(s)} = \mathrm{Si^{+}(s)} + \mathrm{e^{-}(g)} \qquad \bar{I} \tag{5.48}$$

$$\bar{I} = -\bar{\varepsilon}_{\mathrm{v}} + \Delta E_{\mathrm{solv}} = 8.15 + 2.39 - 2.81 = 7.73\,\mathrm{eV} \tag{5.49}$$

This value may be compared to the photoelectric thereshold, I°, which is 5.1 eV. $\bar{I}$ is greater, as it must be, because of the band width broadening.

The second change occurs in the calculation of $\bar{A}$. The value of α in Equation (5.46) was taken as the ionization potential of a free silicon atom, 8.15 eV, in the calculation of $\bar{\varepsilon}_{\mathrm{v}}$. But for $\bar{\varepsilon}_{\mathrm{c}}$ we must take α as 1.39 eV, the electron affinity of a Si atom.

$$\bar{A} = -\bar{\varepsilon}_{\mathrm{c}} - \Delta E_{\mathrm{solv}} = 1.39 - 2.39 + 2.81 = 1.81\,\mathrm{eV} \tag{5.50}$$

This may be compared with $A^{\circ} = 4.0$ eV for Si. It is less positive, as it must be. While we only need about 5 eV of band width to account for the differences, the actual valence band width of silicon is very much larger, about 15 eV. This is a result of the overlappiong of s and p bands.

$(\bar{I} - \bar{A}) = 5.62$ eV is not the same as the spectroscopic gap of $2\beta = 4.78$ eV. The former involves the formation of ions and their solvation, whereas the latter does not create any change in polarity. The spectroscopic gap, which we will call $(I - A)_{\mathrm{s}}$ is the one listed in Table 5.6. It is more closely related to $\bar{E}_{\mathrm{g}}$ than $(\bar{I} - \bar{A})$ is.

Calculation of $\bar{\varepsilon}_{\mathrm{v}}$ and $\bar{\varepsilon}_{\mathrm{c}}$ for the 3–5, 2–6 and 1–7 covalent cases is a little more complicated. From Equation (5.8), we would have

$$\bar{\varepsilon}_{\mathrm{v}} = \frac{n}{8}\alpha_{\mathrm{A}} + \frac{(8-n)}{8}\alpha_{\mathrm{B}} + \frac{[n(8-n)]^{1/2}}{4}\beta \tag{5.51}$$

This is the orbital energy of a bonding MO, ϕ_1, with the composition

$$\phi_1 = \left(\frac{n}{8}\right)^{1/2}\psi_{\mathrm{A}} + \left(\frac{8-n}{8}\right)^{1/2}\psi_{\mathrm{B}} \tag{5.52}$$

where ψ_{A} and ψ_{B} are atomic orbitals on the more metallic element A and the less metallic B, respectively. The anti-bonding MO, ϕ_2, must be

$$\phi_2 = \left(\frac{8-n}{8}\right)^{1/2}\psi_{\mathrm{A}} - \left(\frac{n}{8}\right)^{1/2}\psi_{\mathrm{B}}$$

ϕ_1 keeps the atoms neutral, but ϕ_2 will give an electron distribution so that A is negative and B is positive. We have

$$\bar{\varepsilon}_{\mathrm{c}} = \frac{(8-n)}{8}\alpha_{\mathrm{A}} + \frac{n}{8}\alpha_{\mathrm{B}} - \frac{[n(8-n)]^{1/2}}{4}\beta \tag{5.54}$$

We will take ZnS as an example. The dielectric constant is 5.14, R_0 is 2.34 Å and β is 1.76 eV. Using Equations (5.49) and (5.51), we find

$$\bar{I} = \tfrac{1}{4}(9.39) + \tfrac{3}{4}(10.36) + 1.52 - 1.55 = 10.1\,\text{eV} \tag{5.55}$$

The solvation nergy, 1.55 eV, is calculated for $Zn^{0.25+}$ and $S^{0.75+}$, using the Born equation. With the use of equations (5.50) and (5.54), we calculate

$$\bar{A} = \tfrac{3}{4}(-0.49) + \tfrac{1}{4}(2.08) - 1.52 + 1.55 = 0.2\,\text{eV} \tag{5.56}$$

The solvation energy is for $Zn^{0.75-}$ and $S^{0.25-}$. The numbers are reasonable when compared with $I^\circ = 7.6$ eV and $A^\circ = 4.0$ eV.

Finally, we calculate $(I - A)_s$ by transferring an electron from ϕ_1 to ϕ_2. This is the spectroscopic gap, which should be equal to $\bar{E}_g$.

$$(I - A)_s = -\tfrac{1}{2}(-0.49) + \tfrac{1}{2}(10.36) - 3.04 - 1.24 = 7.2\,\text{eV} \tag{5.57}$$

The solvation energy is now calculated for $Zn^{0.5-}$ and $S^{0.5+}$, which are the net charges after transfer of the electron. By analogous calculations, $(I - A)_s$ has been found for several more examples of covalent compounds. The results are also listed in Table 5.6.

These calculations assume that the charges produced are spread over the entire crystal. Excitons can also be formed, with adjacent charges; the binding energies are much smaller, however. For example, in GaAs the exciton binding energy for $Ga^{0.25-}As^{0.25+}$ would only be 0.33 eV. In the vis–UV spectrum, the exciton absorption is at 1.5 eV, the same as the threshold for conduction band absorption.[47]

In general, the agreement between $(I - A)_s$ and $\bar{E}_g$ is quite good. The numbers need not be the same, as already mentioned. What is important is that the same range of energies is covered and trends are followed. The results for the ionic 2–6 compounds include polarization energies that are larger than for the 1–7 cases by a factor of two or three, bcause the polarizability of O^{2-}, for example, is two to three times larger than for F^-. Actually the polarizability of O^{2-} (and Se^{2-}) is not a constant, but varies with the cation. Note that the excited state of CaO is Ca^+O^-, in a matrix of Ca^{2+} and O^{2-}.

We can also calculate the electronic chemical potential, μ, which is equal to the Fermi energy. From Figure 5.7, we see that there is only one value for μ: that is,

$$\mu = \bar{\mu} = \mu^\circ = -\frac{(\bar{I} + \bar{A})}{2} = -\frac{(I^\circ + A^\circ)}{2} \tag{5.58}$$

The theoretical values for NaCl, Si, ZnS and GaAs are −4.9 eV, −4.8, −5.2 and −4.4, to be compared with the experimental results of −4.8 eV, −4.8, −5.8 and −5.1. Just as in the case of I' and A' measured in solution (see Chapter 3), the solvation effects nearly cancel in the sum of $\bar{I}$ and $\bar{A}$. But they are additive for the difference, and cause large changes.

Assuming that the polarization energies cancel each other, the absolute EN, $\chi = -\mu$, for an ionic solid AB becomes

$$\chi_{AB} = \frac{(I_A + A_B)}{2} = \chi_{A,B} \tag{5.59}$$

This is just the result for a mixture of gas-phase A and B atoms before they react. Also, for the 4–4 cases, we find

$$\chi_{(s)} = \frac{(I + A)}{2} = \chi_{(g)} \tag{5.60}$$

That is, the EN of the solid is the same as the EN of the gaseous atoms which make up the solid!

For the 2–6 and 3–5 covalent solids, $\chi_{(s)}$ is a more complicated mixture of I and A for both atoms, but the final result can be expressed empirically as[48]

$$\chi_{AB(s)} = (\chi_A \chi_B)^{1/2} \tag{5.61}$$

All of these observations can be summed up by saying that there is very little change in the chemical potential in the overall process

$$A(g) + B(g) = AB(s) \qquad - \Delta E_{coh} \tag{5.62}$$

This is quite remarkable, considering the very large change in energy accompanying reaction (5.61).

The hardness, η, on the other hand, almost always increases for reaction (5.62). But this is only true if we take $\bar{E}_g/2 = \eta$. Even so, the hardness *decreases* as the solid is formed from the atoms in the case of Si, Ge and gray tin. The reason for this is not clear.

SOME PROPERTIES OF METALS

The most important fact about the bonding in metals is that it results from a reduction in kinetic energy due to delocalization of the electrons. This is favored by a high coordination number, leading to a high density for most metals. The role of the large number of nearest neighbors is to provide a more uniform potential field in which the electrons move. This keeps the kinetic energy low, in contrast with rapid variations of the potential. The strategy is then to fill the bands of levels only about halfway, thus avoiding the anti-bonding crystal orbitals.

The number of valence orbitals on each atom should be greater than the number of valence electrons, to give the best opportunity to fill only the stable levels. Except in a few cases, it is better to doubly occupy the most stable levels, even though this has an energy cost of increased electron–electron repulsion. The valence atomic orbitals should be rather diffuse, since this will give the greatest overlap, increasing the band width, which means higher delocalization energy.

Such diffuse orbitals are found on atoms of small χ and small η. Diffuse orbitals also overlap at larger atomic distances. This, in turn, permits an increase in the coordination number. Electron–electron repulsion is greatest when both electrons are on the same atom; in this case diffuse orbitals minimize the repulsion energy.[49]

Table 5.7 gives some properties of the more common metallic elements. The cohesive energies are included. These are for a single atom,

$$M(s) = M(g) \qquad \Delta E_{exp} \tag{5.63}$$

so they should be multiplied by two to compare with the cohesive energies of AB compounds in Table 5.3. In general, the metal binding energies are a little less than those of the compounds. Also, the more reactive metals have smaller cohesive energies than the less reactive ones. In a column of the Periodic Table, the binding becomes less on going down for the representative metals, and it goes up for the transition metals.

All of these features are consistent with increasing EN leading to stronger bonds, as expected. However, the noblest and most EN metals, such as Pt and Au, are not as strongly bound as metals such as Ta or W. This is the consequence of the increased number of d-electrons after W, forcing occupancy of anti-bonding orbitals. This is also seen in the first and second transition series.

Let us examine the filling of the crystal orbitals by looking at the first few entries in Table 5.7, in terms of rows of the periodic Table. For the first two rows we have four valence orbitals on each atom. ΔE_{exp} increases for each of the first four electrons to be added in going across the Periodic Table. The bonding orbitals are each doubly occupied. At group 14 (C, Si, Ge) metallic bonding stops, because localized bonding is stronger. After 14, metallic bonding is not found because the extra electrons would have to go into anti-bonding orbitals.

In the third row K uses a doubly occupied 4s band of bonding orbitals, and Ca uses two s–p hybrid orbitals. In Sc the 3d-electron is comparable in energy to the 4s, and the d orbitals form a five-fold degenerate band. The free atoms of the first transition series all have high spin d-electrons in accordance with Hund's rule. We expect similar behavior in the solid state, so that only 2 1/2 electrons will fill the bonding d levels without double occupancy. This is the case for Sc, Ti and V.

For Cr and Mn, the next two electrons must go into anti-bonding levels, causing a drop in ΔE_{exp} as seen in Table 5.7. For Fe and Co the d-electrons go into bonding orbitals with double occupancy, while Ni must divide its extra electron between bonding and anti-bonding. For Cu and Zn, the added electrons are anti-bonding, and the cohesive energy drops. Except for Zn, the transition metals are all paramagnetic, and Fe, Co and Ni are ferromagnetic. The latter phenomenon, along with antiferromagnetism, depends on cooperative effects.

Table 5.7 Some Properties of the Common Metallic Elements

Metal	ΔE_{exp} [kcal/g-atom][(a)]	Φ [eV][(b)]	χ [eV][(c)]
Li	38.6	3.10	3.01
Be	77.9	5.08	4.9
Na	25.9	2.70	2.85
Mg	35.6	3.66	3.75
Al	78.0	4.19	3.23
K	21.4	2.30	2.42
Ca	42.2	2.71	2.2
Sc	90.3	3.50	3.34
Ti	112.6	4.02	3.45
V	122.8	4.44	3.60
Cr	94.9	4.40	3.72
Mn	66.7	3.90	3.72
Fe	99.6	4.65	4.06
Co	101.6	4.70	4.30
Ni	102.8	4.72	4.40
Cu	80.5	4.70	4.48
Zn	31.2	4.30	4.45
Ga	66.5	4.25	3.2
Rb	19.6	2.20	2.34
Sr	39.1	2.76	2.0
Y	101.5	3.5	3.2
Zr	146.0	4.00	3.64
Nb	173	4.20	4.0
Mo	157.5	4.30	3.90
Ru	154	4.80	4.50
Rh	133	4.58	4.30
Pd	90.4	5.00	4.45
Ag	68.1	4.30	4.44
Cd	26.8	4.12	4.33
In	58.0	4.08	3.1
Sn	72.0	4.35	4.30
Sb	62.7	4.56	4.85
Cs	18.7	1.90	2.18
Ba	42.5	2.35	2.40
La	103.0	3.40	3.10
Hf	148	3.65	3.8
Ta	187	4.22	4.11
W	203	4.55	4.40
Re	184	4.95	4.02
Os	189	4.83	4.90
Ir	159	5.05	5.4
Pt	134.9	5.40	5.60
Au	90.5	5.48	5.77
Hg	14.7	4.50	4.91
Tl	43.6	4.02	3.2

Table 5.7 (*continued*)

Metal	ΔE_{exp} [kcal/g-atom][(a)]	Φ [eV][(b)]	χ [eV][(c)]
Pb	46.8	4.18	3.90
Bi	49.5	4.36	4.69

(a) Data from Reference 24.
(b) Data from S. Trassiti, J. Chem. Soc., Faraday Trans. I, *68*, 229 (1972), and Reference 48.
(c) For the free atoms.

The drop in cohesive energy for Zn is exaggerated because of another feature. Increasing nuclear charge draws in the 4s orbital so much that it no longer overlaps so well with its neighbors. This reduces the value of β, the exchange integral, and hence the strength of bonding. For Ga and Ge there is a compensation in that the 4p orbitals are now good bonding orbitals.

For the fourth and fifth rows there are similar variations for the transition metals, except that electron-pairing, or low spin, becomes easier because the orbitals are larger and more diffuse. The poor metallic bonding in mercury is due to an even larger lowering in energy for the 6s orbitals. This is a relativistic effect. Near the nucleus, the 6s-electron is travelling nearly at the speed of light. Its mass increases and the heavier electron is drawn closer to the nucleus. This is the cause of the so-called inert-pair effect, which dominates the chemistry of the fifth row.[50]

The energy level diagram for metals is much simplified, since the disappearance of the energy gap of Figure 5.7 causes μ, ε_F, $-I^\circ$ and $-A^\circ$ to coalesce to the same level. Figure 5.8 shows this and also includes an average energy level for

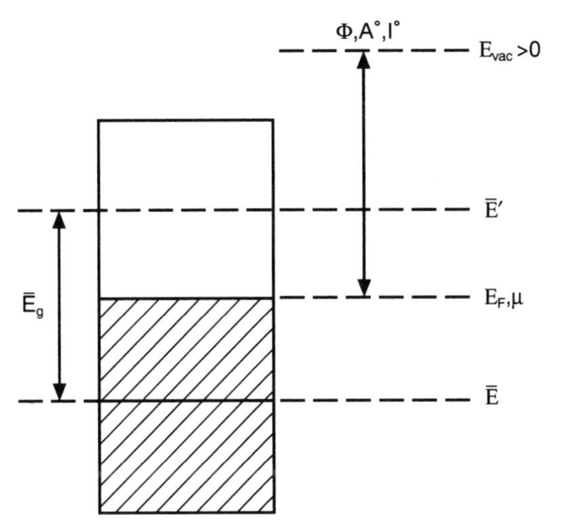

Figure 5.8 Energy levels in a metal; $\bar{\varepsilon}$ and $\bar{\varepsilon}'$ are the average energies of the filled and unfilled levels, respectively

both the filled and unfilled levels of the conduction band. The average energy gap between them cannot be determined by the use of Equation (5.40), since the dielectric constant of a metal is infinite.

Also, for metals we cannot find $(I - A)_s$ by examining the electronic spectrum. Because of the band structure, all possible frequencies of the electromagnetic spectrum can be absorbed. This is usually followed by the immediate re-emission of the photon so that there is almost total reflectivity.[51] In the visible, this accounts for the appearance of metallic luster. Above the photoelectronic threshold, electrons are emitted from the metal as well.

Even though the bonding in metals must be purely covalent, we cannot use the simplified bonding model of the earlier section. That model is appropriate for cases where the delocalized crystal orbitals can be replaced by average localized orbitals. This is not possible for metals, or at least not easy. Actually the tight binding theory at the Hückel level of approximation has been used for metals in several cases.[52]

It is necessary to distinguish between s, p and d orbitals, and to use different exchange integrals for σ-, π- and δ-bonding. If this is done, one can successfully account for differences in energy. For example, the choice of crystal structure (fcc, hcp or bcc) for different metals can be predicted.[53] It is likely that cohesive energies could also be calculated in this way, if values of β such as those in Table 5.2 were used. Unfortunately, most of the β values listed for metals were estimated from the experimental cohesive energies.

Assuming that the cohesive energy is due to the delocalization energy of partly filled bands, we can conclude that Equations (5.20), (5.22) and (5.23) are valid. We can also assume that the coulombic energy α is the same as for the free metal atoms. This is what is done in the simple Hückel theory used for other covalent bonding. Then we can write

$$\Delta E_{\text{exp}} = Nn\beta \tag{5.64}$$

where N is the number of valence electrons per atom, and n is an unknown factor.

In the case of the group 1 and 11 metals we can use the β of Table 5.2. Then Equation (5.64) gives $n = 3.1$ for the alkali metals and $n = 5$ for Cu, Ag and Au. These are reasonable values in a band theory of metal bonding. For other metals, we simply lump together the unknown n and β. We can also use the properties of a band to write

$$\bar{\varepsilon} = \alpha + n\beta \qquad \text{filled levels} \tag{5.65}$$

$$\bar{\varepsilon}' = \alpha' - n\beta \qquad \text{empty levels} \tag{5.66}$$

For a half-filled band, $\varepsilon_F = \alpha = \mu$.

The other experimental property which we have available is the work function, Φ. According to Equation (5.31) this is equal to μ, except for the correction due to surface potentials. Table 5.7 shows the experimental results for the work functions of the common metals. They are usually obtained for polycrystalline samples, since the surface potentials vary according to the crystal face through which the electron is emitted.

We see at once that Φ is not equal to the ionization potential of the free atom, as we might have supposed. Instead it is very nearly equal to the electronic chemical potential, μ, of the atom. These data are also listed in Table 5.7 for convenience. The agreement is really quite good. This should not surprise us, since we know that the coulombic integral can have two values. One, for the removal of an electron is $-I$. The other, for the addition of an electron, is $-A$.

Since the energy levels in the band theory must be continuous, the value $\varepsilon_F = -(I + A)/2$ is a reasonable result. We now have $-\mu = (I^\circ + A^\circ)/2 = (I + A)/2$, where $I^\circ = A^\circ = -\varepsilon_F$ and I and A are for the free atom. The fact that μ hardly changes in going from the atoms to the solid is the same result that we found for the other classes of solids.

To find the spectroscopic gap, $(I + A)_s$, we can assume that $\alpha = \alpha'$ since there is no change in the charge. Then

$$\bar{E}_g = -2n\beta = 2\Delta E_{exp}/N \tag{5.67}$$

This result is very similar to that for other covalent solids, and obviously gives us a direct dependence of the cohesive energy on the effecive energy gap, or hardness $= \bar{E}_g/2$.

In accordance with the standard operational definition of the hardness, we must have $\eta = (I^\circ - A^\circ)/2 = 0$. This agrees with the ease of electron movement, as evidenced by the high electrical and thermal conductivity. However, this movement does not lead to the breaking of bonds. The energy gap, $\bar{E}_g$, is obviously the quantity which determines chemical reactivity, such as dissociation. The hardness decreases in going from the free atoms to the solid, if we take $(I^\circ - A^\circ)/2$. But it increases if we take $\bar{E}_g/2 = \Delta E_{exp}/N$ as an effective hardness.

The relation between the work function and the Mulliken electronegativity, as shown in Table 5.7, has been known for some time.[48,54] The early use of this near-equality was to determine the unknown EN of some of the elements. Now we would turn the procedure around and estimate Φ from the usually well-known data for the free atoms. In any case, the most interesting point is the constancy of the electronic chemical potential.

It is not easy to find other data which show that $\bar{E}_g$ plays a role in the reactions of solids. There is one case for metals which can be used as a test: the formation of alloys, which represents, at least in a sense, the reaction of two solid metals with each other. The stability of alloys of varying composition is influenced by factors such as the relative atom sizes, and the number of valence electrons per atom (the Hume–Rothery rules).

But it has long been known that differences in EN for two metals are favorable for stable alloy formation. This has been put in a quantitative form by Miedema and his co-workers.[55] The heat of formation of a binary alloy can be given by the empirical equation

$$\Delta H = f(c)[-P(\Delta\phi)^2 + Q\Delta\rho_s^{2/3}] \tag{5.68}$$

where $f(c)$ is a composition function P and Q are constants for all mixtures, $\Delta\Phi$ is the difference in work functions for the two components, and $\Delta\rho_s$ is the difference in the electron density at the surfaces where the atoms of the two kinds meet. It is positive and gives the work needed to make the electron density continuous at these surfaces.

The term $-P(\Delta\Phi)^2$ is the one which makes the alloy stable. The similarity to the DFT equation

$$\Delta E = -\frac{(\mu_C - \mu_D)^2}{4(\eta_C + \eta_D)} \tag{5.69}$$

is apparent. P is equal to about $1\,eV^{-1}$ and would be equal to $0.25(\eta_C + \eta_D)^{-1}$, if Equation (5.69) is assumed to be operative. We see that $(\eta_C + \eta_D)$ would be equal to 0.25 eV, if this were so. Table 5.7 shows that $(\eta_C + \eta_D)$ would be 5–10, if $\bar{E}_g/2$ were the hardness, and would vary with each pair of metals. The experimental value of P is closer to zero than to $\bar{E}_g/2$, but it is not clear what determines it. In any case, Equation (5.68) is remarkably accurate in predicting the possibility of forming a stable alloy.

CLUSTERS AND SURFACES

These two topics are treated together because they may both be considered as dealing with solids that are incomplete in some way. Clusters may be as small as three atoms or may contain many thousands of atoms. The study of clusters, particularly metal clusters, is a very active field.[56] The interest is two-fold: one purpose is to learn how the properties of solids emerge as the number of units increases to infinity; the other is because clusters are important in heterogeneous catalysis. They have very large ratios of surface area to mass and are much more reactive than large crystals.

For orientation, let us see how some important properties change as we go from diatomic molecules to solids, for several kinds of systems. This is illustrated in Table 5.8.

The largest change is in the cohesive energy, which increases in all cases. For ionic solids, the increase is due to the larger Madelung constant. For metals, the increase is due to the better delocalization energy with a higher coordination number. For covalent solids, the increase occurs because all of the valence-shell

Table 5.8 Some Properties of Diatomic Molecules and Solids

	ΔE_{coh} [kcal/mol]	I [eV]	A [eV]
NaCl(g)	98	8.9	0.8
NaCl(s)	153	9.0	0.5
Li_2(g)	24	5.1	0.4
Li(s)	77[a]	3.1	3.1
Si_2(g)	74	7.4	2.0
Si(s)	219[a]	5.1	4.0
GaAs(g)	50	7.0[b]	0.5[b]
GaAs(s)	156	5.4	4.0

[a] For two atoms.
[b] Calculated for Ga_2As_2. From Reference 57.

orbitals and electrons can be utilized. For ionic solids, there is little change in I or A, because the band width is small. For metals and covalent solids, I becomes smaller and A becomes larger because of the band width effect.

As the cluster size increases, two kinds of behavior are seen. In some cases ΔE_{coh}, I and A all change rather smoothly as the number of units in the cluster grows. An example would be nickel clusters, Ni_N, where N is the number of atoms.[58] The quantity to compare for energy is BE/N where BE is the total bonding energy. This number is 1.10 eV or Ni_2, 2.70 eV for Ni_{13}, 4.06 eV for Ni_{147} and 4.44 eV for nickel metal. At the same time, I decreases and A increases rather regularly, so that they approach one another.

This behavior for I and A can be predicted on classical grounds. The work function for bulk metal would be modified for small spherical samples by the Born charging energy. The ionization potential would be increased and the electron affinity would be decreased by the same amount

$$I = \Phi + e^2/2R \tag{5.70}$$

$$A = \Phi - e^2/2R \tag{5.71}$$

where R is the radius of the sphere. As R goes to infinity, I and A become equal, and equal to the work function. Equations similar to (5.70) and (5.71) have been derived, both classically and quantum mechanically, except that values slightly different from 1/2 are predicted.[59]

The other kind of general behavior observed for both metallic and non-metallic clusters, is that ΔE_{coh}, I and A all change in a non-uniform way consistent with the existence of shell structure within the cluster. That is, there is evidence for extra stability for certain values of N, called magic numbers.[60] The evidence often comprises increased intensity in the mass spectra for the magic numbers. This has been seen for the alkali metals and the noble metals of Group 11.

Another kind of evidence is a sudden drop in ionization potential as N is varied. This has been noted for carbon clusters.[61] The interpretation is that the previous value of N corresponded to a filled shell and that the added electron had to go into a higher-energy shell. The best evidence would be a measurement of the bonding energies of the various clusters. This is difficult experimentally, but a number of theoretical calculations have been made.

The procedure is to solve the wave equation for electrons in a suitable spherically symmetric positive potential.[62] The solutions depend on quantum numbers, much like those of atoms, and even more like those of nuclei. The magic numbers correspond to the filling of shells. For the alkali metals the magic numbers are $N = 2$, 8, (18,) 20, 34, 40 and 58.[60]

Figures 5.9 and 5.10 show the results of some theoretical calculations on lithium clusters using the spherical jellium background model.[63] In this model the ionic charge is spread uniformly over a sphere of radius R, which is proportional to $N^{1/3}$. The calculations were made for Li_N, Li_N^+ and Li_N^-, with N increasing from 2 to 67. Total energies were found.

The cluster stability is shown by plotting the second energy difference

$$\Delta_2 = E(N+1) + E(N-1) - 2E(N) \tag{5.72}$$

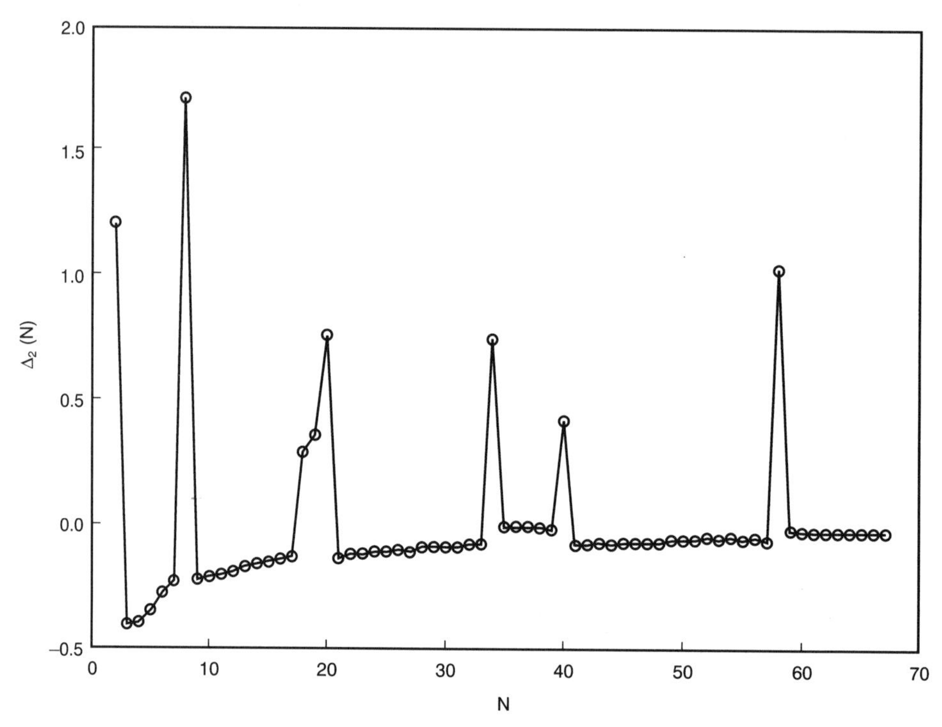

Figure 5.9 Second energy differences, Δ_2, for lithium clusters versus the number of atoms, N Reprinted with permission from Reference 63

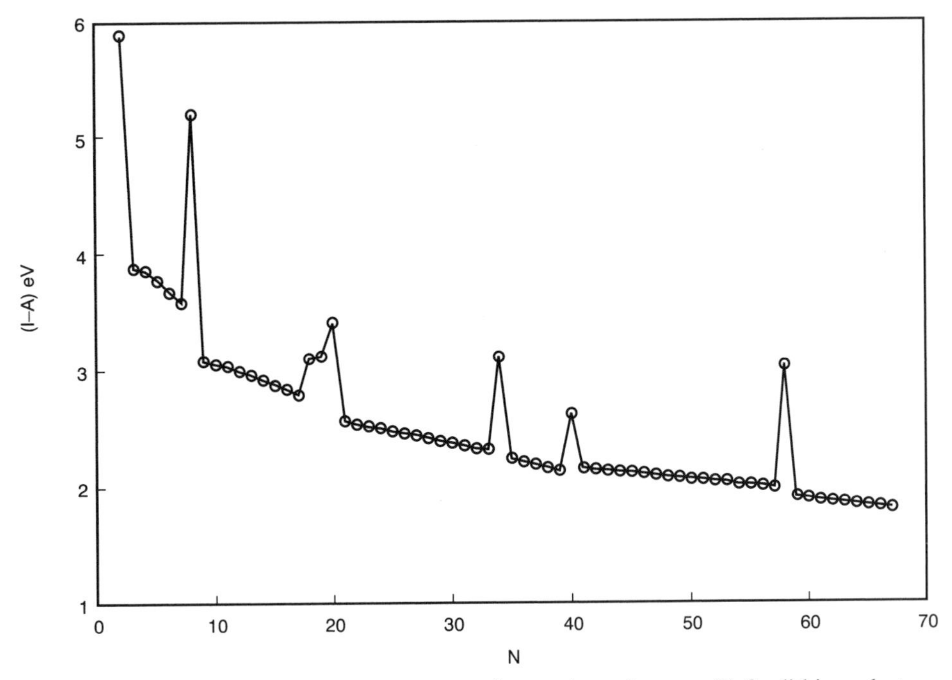

Figure 5.10 Chemical hardness, $I - A$, versus the number of atoms, N, for lithium clusters. Reprinted with permission from Reference 63

Figure 5.9 shows that Δ_2 is close to 0, except when $N = 2, 8, 18, 20, 34, 40$ and 58. The peaks at these numbers show that the model is a good one. Figure 5.10 shows a plot of $(I - A)$ as a function of N. There is a general decrease of $(I - A)$ as N increases. This is consistent with Equations (5.70) and (5.71).

However, the most striking features of Figure 5.10 are the pronounced peaks at $N = 2, 8, 18, 20, 34, 40$ and 58. At the magic numbers the hardness shows a local maximum. Increased stability is accompanied by increased hardness. Identical results have been calculated for sodium clusters.[59] Similar, but not so spectacular, resuls may be calculated for small silicon clusters.[64]

The case of carbon clusters is quite different.[65] A model suitable for metals will not work. Instead, a definite structure must exist using ordinary covalent bonds. For $N < 30$ the clusters exist in linear, cyclic and bicyclic forms.[66] For $N \geq 30$ the structures are three-dimensional (3D) and evolve into the cage or soccer-ball structure characterisic of C_{60} and C_{70}. For $N \leq 30$, the magic numbers are given by $N = 4n + 2$, where n is an integer.[61] This suggests an aromatic type of stabilization.

For larger N the magic numbers are 50, *60*, *70*, 90, 100.[67] The criterion for extra stability is that the carbon atoms can form a 3D structure in which the MOs are such that all N electrons are in bonding orbitals, and all non-bonding orbitals are empty.[68] In other words, there must be a large HOMO–LUMO gap.

There is another form of carbon clustering whose evolution from monomer to solid is of interest: this is the condensation of benzene rings to graphite. A regularity is found in the changes of χ and η (Table 5.9). The near-constancy of χ all the way to graphite, where χ is also the work function, was noted by Becker and Wentworth in an interesting early paper.[69] This work was one of the first times that the term "molecular electronegativity" was used.

At the same time, the hardness is decreasing steadily. Graphite is a good conductor, albeit not an isotropic one. The band gap is near zero for the x–y plane. In Chapter 2 it was noted that decreasing η meant increasing reactivity, or less REPE (resonance energy per electron), as the number of rings increased. This does not mean that graphite is extremely reactive, any more than does the zero gap for metals. In fact graphite has an REPE of 4.7 kcal/mol, almost as much as the 6.0 kcal/mol of benzene.[70]

Surfaces are incomplete solids, in that the surface atoms have no nearest neighbors in one of the six Cartesian directions. This means that there are "dangling" atomic orbitals, both filled and empty, which are not being used. This applies, of course, to a clean surface. Which atomic orbitals are at the surface, and their orientation, depend on an arbitrary choice for a coordinate system and on which crystal plane forms the surface. In any case, the atomic orbitals form linear combinations called surface orbitals. The interaction is usually weaker than in the bulk of the solid.

Some general remarks can be made about the properties of surfaces without knowing the detailed structures. First of all, surface atoms are a a higher energy than bulk atoms. This leads to surface tension, which is the excess energy needed to form a unit area of surface. Secondly, the electronic chemical potential for surface atoms must be the same as for the bulk atoms. However, the work function for the surface atoms will be different from that of bulk atoms.

It is difficult to estimate the difference. Since the surface atoms are at a higher energy, and this means higher *electronic* energy, one would expect that it would be easier to remove an electron from the surface atoms. But there is an opposing effect. In forming the crystal orbitals the contribution of the surface orbitals will be different from that of bulk orbitals. General MO theory tells us that

Table 5.9 Values of χ and η for Benzene, Various Condensed Carbocyclic Compounds and Graphite

	χ [eV]	η [eV]
C_6H_6	4.1	5.3
$C_{10}H_8$	4.0	4.2
$C_{14}H_{10}$	4.0	3.5
$C_{18}H_{12}$	4.1	3.4
Graphite	4.4	0.0

they will contribute less to the more stable crystal orbitals, but more to the less stable ones.

Therefore the surface will develop a positive charge when the valence band is less than half-full, but will become more negative as the band fills. The positive charge will counteract the higher surface energy and make the work function greater. Filled bands, however, should have a smaller work function. There is experimental evidence for this variation.[71]

There are other problems with trying to predict the properties of surfaces. In an effort to reduce the surface tension, the atoms of the surface will usually rearrange themselves to some degree. The changes can be small, such as a variation in bond lengths, or extensive, leading to a quite different structure for surface atoms. In this case the surface is said to have been reconstructed. Moreover, real surfaces are not the uniform flat planes that we visualize, but contain many steps, kinks and defects.[72]

In an attempt to lower their surface energy, solids almost always will adsorb small molecules, such as H_2O or O_2, on their surfaces. Molecules which form molecular solids have much smaller surface tensions than other solids. The study of surfaces is a difficult one, it can be seen. Fortunately a number of very special experimental methods have been developed,[72,73] and surfaces may be studied in great detail. Such studies are useful in many important practical areas, such as adhesion, lubrication, corrosion and adsorption. However, the most important area is probably heterogeneous catalysis.

This subject is too vast to cover even superficially, but it is worthwhile to show how the modern theory of heterogeneous catalysis is related to the topics in this book.[74] The key reaction is between the adsorbate molecule, acting as one reactant, and the surface atoms of the catalyst, acting as the second reactant. There is a transfer of electron density between the two reactants leading to chemisorption, and bond-breaking or weakening in the adsorbate.[71,75] The acid–base character of the surface is matched to that of the adsorbate.[76]

Donating and accepting orbitals are identified and their overlaps considered. The direction of electron flow is found by considering the work function of the solid and I and A for the gas-phase adsorbate molecule. Until recently the DFT-based concepts of absolute EN and hardness were not used. The older idea of HSAB was used in cases where acid–base interaction seemed important.[76]

The most common theoretical tool used is calculation by the extended Hückel theory (EHT).[77] For solids these lead to a density-of-states (DOS) diagram. This is a picture of the number of energy levels per unit of energy, as a function of the energy. A high density of states at the Fermi level leads to stronger adsorption of a substrate. It may also be recalled that the DOS at the Fermi level is equal to the local softness.[78] In DFT a high value of the softness also leads to better interaction with a substrate.

The examples of most interest are the reactions of the small molecules H_2, N_2, CH_4, O_2 and CO on transition-metal surfaces. It was concluded in these important cases that the main interaction was the donation of metal d-electron density into the σ^* and π^* orbitals of the substrate. It was also found that, in any transition series, the early members such as Ti, Zn or Ta were better donors than the late members, such as Ni, Pd or Pt. This does not mean that the former were better catalysts, since too strong an interaction will reduce catalytic efficiency.

We can also consider these interactions from the viewpoint of DFT. Hopefully the equation

$$\Delta N = \frac{(\mu_D - \mu_C)}{2(\eta_D + \eta_C)} = \frac{(\chi_C - \chi_D)}{2(\eta_D + \eta_C)} \tag{5.73}$$

will show a correlation between the amount of electron transfer and the strength of bonding. For metals the work function will be used for χ_D and a value of ~ 0 for η_D. The very large number of bulk atoms will serve as a reservoir for the few surface atoms that actually react.

For the reactions of H_2, N_2, O_2 and CO we would reach the same conclusions as the extended Hückel calculations. These are molecules of high EN. However, for CH_4 the predicted direction of net electron flow would be from CH_4 to any of the transition metals. In the extended Hückel approach, there is electron transfer in both directions, with a small preference for metal d to σ^*. Note that the reaction of methane with metal surfaces is much more difficult than for the other substrates.

Equation (5.73) gives the same order of strength of bonding for the various metals, since Φ and χ become larger as we go from left to right in a transition series, both for the free atoms and the bulk metals. The near-equality of χ and Φ in table 5.7 suggests that the relative reactivity of various substrates should be the same for the atoms and their metals. The ordering for free atoms is given in Table 3.6 of Chapter 3. It seems to be reasonable for metals also, though it is hard to find comparable data.

RECENT APPLICATIONS OF CONCEPTS

The use of the DFT-based concepts in solid-state chemistry is in its infancy. But enough has been done to show promise in a number of diverse areas. An important application appears in the interpretation of scanning tunneling microscopy (STM) signals.[79] STM, invented in the early 1980s by Binnig and Rohrer, is one of the ingenious new techniques for studying surfaces. It generates images of the atoms on a surface by the tunneling of electrons between the atoms and a sharp metal tip. The current can flow either way, depending on the bias voltage. The brightness of the image depends on the magnitude of the current.

At low temperature and voltage, the current is directly proportional to the local density of states at each point on the surface.[80] But this DOS is just equal to the local softness! Therefore it is possible to scan a surface to find the sites of greatest softness, and hence of reactivity.[79] Both the original surface and molecules adsorbed on the surface can be examined. By changing the bias voltage, the local softness for both accepting and donating electrons can be measured.

The current and the brightness of the signal also depend on the work function of the surface.[80] Since this could change for various points on the surface, it may also play a role. However, there is already some evidence pointing to a major role for the local softness. The (111) plane of silicon contains two kinds of Si atoms, one kind being softer for accepting electrons. Soft molecules are adsorbed on the softer atoms and hard molecules on the harder atoms.[79] Organic molecules adsorbed on surfaces show bright spots for aromatic rings but are dark for aliphatic chains. The brightness of functional groups increases in the order $CH_3 \simeq OH \simeq Cl < NH_2 < Br < I < SH$.[81]

The HSAB principle has long been used to rationalize the adsorption of various molecules on various surfaces (see Chapter 1). This correlation is now being done more quantitatively. For example, the adsorption of C_6H_5SH and $C_6H_5SO_2H$, and the non-adsorption of $C_6H_5SO_3H$ on gold, have been explained in terms of the increasing HOMO–LUMO gap.[82] The sites of adsorption of Na, Al, As and Cl on a gallium arsenide surface agree with predictions based on local softness.[57] The HOMO–LUMO gap and its role in the adsorption of organic polymers on metals has also been discussed.[83] This is important in understanding the operation of adhesives.

There is a theory explaining the stronger adsorption of H_2 on the early transition metals of each series, and the non-adsorption on Au, Ag and Cu.[84] As the H_2 molecule approaches the metal surface, the electron in the s orbital of the metal is pushed into an empty d orbital to avoid repulsion. This detailed mechanism has been confirmed by DFT calculations.[85] The local DOS as well as orbital symmetry are the determinants. These also determine the surface sites where a hydrogen atom will be bound.

A rather remarkable paper has appeared on catalytic reactions in zeolites.[86] It combines *ab-initio* calculations on zeolites with the results of the following reaction, catalyzed by the same zeolites

$$C_6H_5CH_3 + CH_3OH \rightarrow C_6H_4(CH_3)_2 + H_2O \quad (5.74)$$

Both *ortho*- and *para*-xylenes are formed as products. In toluene, the *ortho* position is more negative than the *para*, but the local softness is greater as the *para* position. If coulombic effects are dominant, we expect a low *p*/*o* ratio. If electron transfer, or orbital control, is dominant, we should have a high *p*/*o* ratio.

Both *ab-initio* and semi-empirical MO calculations were made for model clusters of zeolites, in which Al/Si and Al/Ga/B ratios were changed. The HOMO–LUMO gap was found in each case. The experimental *p*/*o* ratio, indeed, was found to vary inversely with the size of the gap. Softer zeolites favored orbital control, leading to more *para*-xylene, as would be predicted.

There have been several studies in which certain properties of crystalline compounds have been calculated. These include charges on the atoms, acidity, basicity and reactivity of atoms and bonds. The approaches are based on absolute EN and local softness, primarily. The objectives are to estimate the strength of various bonds, reactivity of sites to electron donors and acceptors, and so on. It is to be hoped that such information can be useful in synthesis of complex substances and their stabiliy to various internal disproportionation changes. Both electronegativity equalization methods (EEM)[87] and EHT calculations[88] have been used.

Most real solids are polycrystalline, consisting of grains that are more or less cemented together at their boundaries. Impurities may or may not segregate at these grain boundaries. This in turn has a large effect on materials used in electronic devices. The HSAB and Maximum Hardness Principles have recently been used to study segregation of As and Ga in germanium crystalites.[89]

Ab-initio calculations were made on combinations of Ge and As, with As in the bulk and with As on the surface. The local softness, both for accepting electrons and donating electrons, was calculated. It was found that arsenic, a soft impurity (shallow donor) did congregate at the surface. A hard impurity (deep donor) was predicted not to. Also it was predicted that gallium (a soft acceptor) would not segregate at the grain boundaries. The PMH was obeyed for each kind of behavior.

An interesting example of the hardness concept has been given by a study of the charge capacity of TiS_2 intercalated with lithium.[90] It may be recalled that Huheey had originally called $(I - A)^{-1}$ the charge capacitance of an atom or group, κ.[91] This can be written in terms of the electronic chemical potential and the charge, Q, transferred to the group,[92]

$$\kappa = \frac{1}{2\eta} = \frac{Q}{(\mu^\circ - \mu)} \tag{5.75}$$

where μ° is the chemical potential before electron transfer.

The energies of the HOMO and LUMO of a Ti_8S_{32} cluster were calculated by the EHT, as a function of the amount of Li intercalcated between two layers of Ti_4S_{16}. The role of the Li was to transfer charge to the TiS_2 layers. It was found that the experimental voltage–composition curve could be reproduced.

REFERENCES

1. J.C. Slater, *Phys. Rev.*, **81**, 385 (1951).
2. P.A.M. Dirac, *Proc. Cambridge Philos. Soc.*, **26**, 376 (1930).
3. J. Callaway and N.H. March, *Solid State Physics*, **38**, 135 (1984).

4. J.C. Slater, *The Self-Consistent Field Method for Molecules and Solids*, McGraw-Hill, New York, Vol. 4, 1974.
5. W. Yang, Phys. Rev. Lett., *66*, 1438 (1991); idem, *Phys. Rev. A*, **44**, 7823 (1991).
6. R.G. Pearson, *J. Mol. Struct. (Theochem.)*, **260**, 11 (1992).
7. C.A. Coulson, *Valence*, 1st Edn., Clarendon Press, Oxford, 1952, p. 262ff.
8. R.G. Pearson and H.B. Gray, *Inorg. Chem.*, **2**, 358 (1963); G. Klopman, *J. Am. Chem. Soc.*, **86**, 4550 (1964); R.L. Flurry, *J. Phys. Chem.*, **69**, 1927 (1965); idem, ibid., **73**, 2111 (1969).
9. For introductory reviews, see M. Kertesz, *Int. Rev. Phys. Chem.*, **4**, 125 (1985); J.A. Duffy, *Bonding, Energy Level and Bands*, Longmans, New York, 1990.
10. W.A. Harrison, *Electronic Structures and Properties of Solids*, Freeman, San Francisco, 1980.
11. For reviews and analyses, see P.G. Nelson, *J. Chem. Ed.*, **71**, 24 (1994); J. Meister and W.H.E. Schwarz, *J. Phys. Chem.*, **98**, 8245 (1994).
12. R.G. Pearson, *Inorg. Chem.*, **30**, 2856 (1991).
13. J.K. Burdett, *J. Am. Chem. Soc.*, **102**, 450 (1980).
14. W.B. Pearson, *J. Phys. Chem. Solids*, **23**, 103 (1962); J.C. Phillips, *Rev. Mod. Phys.*, **42**, 317 (1970).
15. R.S. Mulliken, *J. Am. Chem. Soc.*, **72**, 4493 (1950); idem, *J. Phys. Chem.*, **56**, 295 (1952).
16. N. Born and K. Huang, *Dynamical Theory of Crystal Lattices*, Oxford University Press, London, 1954.
17. W. Yang and R.G. Parr, *Phys. Chem. Miner.*, **15**, 191 (1987); G. Simmons and H. Wang, *Single Crystal Elastic Constants*, MIT Press, Cambridge, MA, 1971.
18. H. Wang, *Phys. Chem. Miner.*, **3**, 251 (1978).
19. C.A. Coulson, L.B. Redel and D. Stocker, *Proc. R. Soc. London*, **270**, 357 (1962).
20. M. Wolfsberg and L.J. Helmholz, *J. Chem. Phys.*, **20**, 837 (1952).
21. S.G. Bratsch, *Polyhedron*, **7**, 1677 (1988).
22. P. Politzer, *J. Am. Chem. Soc.*, **91**, 6235 (1969); idem, *Inorg. Chem.*, **16**, 3350 (1977).
23. R.W.G. Wyckoff, *Crystal Structures*, 2nd Edn., Vol. 1, Interscience, New York, 1963; P. Villars and J.D. Calvert (Eds.), *Pearson's Handbook for Crystal Data*, American Society of Metals, Metals Park, OH, 1985.
24. D.D. Wagman, W.H. Evans, V.B. Parker, R.H. Schumm, I. Halo, S.M. Bailey, K.L. Churney and R.L. Nuttall, "The NBS tables of chemical thermodynamic properties", *J. Phys. Chem. Ref. Data*, **11**, Suppl. 2 (1982).
25. J. Simons and A.I. Moldyrev, *J. Phys. Chem.*, **100**, 8023 (1996).
26. For a simple introduction, see R. Hoffmann, *Solids and Surfaces*, VCH Weinheim, 1988.
27. For a more detailed account, see J.K. Burdett, *Chemical Bonding in Solids*, Oxford University Press, New York, 1995.
28. For example, see J.R. Hook and H.E. Hall, *Solid State Physics*, 2nd Edn., John Wiley and Sons, New York, 1991, Chapter 3.
29. T.A. Albright, J.K. Burdett and M.H. Whangbo, *Orbital Interactions in Chemistry*, Wiley-Interscience, New York, 1985.
30. Reference 27, p. 96ff.
31. N.W. Ashcroft and N.D. Mermin, *Solid State Physics*, Saunders, New York, 1976.
32. For example, see J. Goodisman, *Electrochemistry: Theoretical Foundations*, Wiley-Interscience, New York, 1987.
33. H. Reiss and A. Heller, *J. Phys. Chem.*, **89**, 4207 (1985).
34. For a brief review see Reference 3.

35. For example see M. Causà and A. Zupan, *Chem. Phys. Lett.*, **220**, 145 (1994); idem, *Int. J. Quantum Chem.*, **56**, 337 (1995).
36. J.R. Christman, *Fundamentals of Solid State Physics*, John Wiley and Sons, New York, 1988, Chapter 8.
37. R.G. Parr and W. Yang, *Density Functional Theory of Atoms and Molecules*, Oxford University Press, New York, 1989, p. 77.
38. For example, see Reference 28, Chapter 6, and Reference 36, Chapter 14.
39. (a) D.R. Penn, *Phys. Rev.*, **1311**, 5179 (1978); (b) J.A. van Vechten, *Phys. Rev.*, **182**, 891 (1969); idem, ibid., **187**, 1007 (1969).
40. J.C. Phillips and J.A. van Vechten, *Phys. Rev. Lett.*, **22**, 705 (1969); J.C. Phillips, *Rev. Mod. Phys.*, **42**, 317 (1971).
41. J.K. Burdett, B.A. Coddens and G.V. Kulkarni, *Inorg. Chem.*, **27**, 3259 (1988).
42. N.F. Mott and M.J. Littleton, *Trans. Faraday Soc.*, **34**, 485 (1938).
43. R.T. Poole, J.G. Jenkins, J. Leisegang and R.C.G. Leckey, *Phys. Rev.*, **1311**, 5179 (1975).
44. See Reference 4, Chapter 6.
45. E. Burstein, A. Pinczuk and R.F. Wallis, *The Physics of Semimetals*, D.L. Carter and R.T. Bate, Eds., Pergaon Press, New York, 1971, p. 251.
46. M. Aniya, *J. Chem. Phys.*, **96**, 2054 (1992).
47. M.D. Sturge, *Phys. Rev.*, **127**, 768 (1962).
48. E.C.M. Chen, W.E. Wentworth and J.A. Ayala, *J. Chem. Phys.*, **67**, 2842 (1977).
49. For a discussion of the boundary between metals and nonmetals, both in the elements and in compounds, see P.P. Edwards and M.J. Sienko, *Acc. Chem. Res.*, **15**, 87 (1982).
50. K.S. Pitzer, *Acc. Chem. Res.*, **12**, 271 (1979); P. Pykko and P. Declaux, ibid., 276.
51. P.A. Cox, *The Electronic Structure and Chemistry of Solids*, Oxford University Press, Oxford, 1987, Chapters 2 and 7.
52. See Reference 27, Chapters 2 and 4, for some examples.
53. J.K. Burdett and S. Lee, *J. Am. Chem. Soc.*, **108**, 3050, 3063 (1985).
54. R.T. Poole, D.R. Williams, J.D. Ridley, J.G. Jenkins, J. Liesegang and R.C.G. Leckey, *Chem. Phys. Lett.*, **36**, 401 (1975).
55. A.R. Miedema, F.R. DeBoer, *J. Phys. F.*, **3**, 1558 (1973); R. Boom, F.R. DeBoer and A.R. Miedema, *J. Less Common Metals*, **46**, 271 (1976).
56. For reviews, see J. Koutecky and P. Fantucci, *Chem. Rev.*, **86**, 539 (1986); J.C. Phillips, ibid., **86** 619 (1986). M.D.Morse, ibid., **86**, 1049 (1986).
57. P. Piquini, A. Fazzio and A. DalPino, Jr., *Surf. Sci.*, **313**, 41 (1994).
58. G. Pacchioni, S.C. Chung, S. Kruger and N. Rösch, *Chem. Physics*, **184**, 125 (1994).
59. For a discussion, see J.A. Alonso and L.C. Balbas, *Structure and Bonding*, **80**, 229 (1993).
60. W.A. de Heer, W.D. Knight, M.Y. Chou and M.L. Cohen, *Solid State Physics*, **40**, 93 (1987).
61. S.B.H. Bach and J.R. Eyler, *J. Chem. Phys.*, **92**, 358 (1990).
62. T.P. Martin, T. Bergmann, H. Göhlich and T. Lange, *J. Phys. Chem.*, **95**, 6421 (1991).
63. M.K. Harbola, *Proc. Natl. Acad. Sci. USA*, **89**, 1036 (1992).
64. K. Rhagavachari and C.M. Rohlfing, *J. Chem. Phys.*, **94**, 3670 (1991); W. von Niessen and V.G. Zahrzewski, ibid., **98**, 1275 (1993).
65. For a review, see G.S. Hammond and V.J. Kuck, (Eds.), *Fullerenes. Synthesis, Properties and Chemistry*, American Chemical Society, Washington, DC, 1992.
66. G. von Helding, M.T. Hsu, P.R. Kemper and M.T. Bowers, *J. Chem. Phys.*, **95**, 3835 (1991).
67. J.A. Zimmerman, J.R. Eyler, S.B.H. Bach and S.W. McElvany, *J. Chem. Phys.*, **94**, 3556 (1991).

68. N. Kurita, K. Kobayashi, H. Kumahora, K. Tago and K. Ozawa, *Chem. Phys. Lett.*, **188**, 181 (1992).
69. R.S. Becker and W.E Wentworth, *J. Am. Chem. Soc.*, **85**, 2210 (1963); R.S. Becker and E. Chen, *J. Chem. Phys.*, **45**, 2403 (1966).
70. C.A. Coulson and R. Taylor, *Proc. Phys. Soc. A*, **65**, 815 (1952).
71. (a) E. Shustorovich and R.C. Baetzold, *J. Am. Chem. Soc.*, **102**, 5989 (1980); (b) J.Y. Saillard and R. Hoffmann, ibid., **106**, 2006 (1984).
72. For an excellent review of surface phenomena of all kinds, see G.A. Somorjai, *Chemistry in Two Dimensions*, Cornell University Press, Ithaca, New York, 1981.
73. R.J. Hamers, *J. Phys. Chem.*, **100**, 13103 (1996).
74. For introductions, see R. Hoffmann, *Solids and Surfaces*, VCH, New York, 1988; G.A. Somorjaj, *Introduction to Surface Chemistry and Catalysis*, Wiley, New York, 1994; D.W. Goodman, *J. Phys. Chem.*, **100**, 13090 (1996).
75. (a) R.C. Baetzold, *J. Am. Chem. Soc.*, **105**, 4271 (1983); H.S.S. Sung and R. Hoffmann, ibid., **107**, 578 (1985).
76. P.C. Stair, *J. Am. Chem. Soc.*, **104**, 4044 (1982).
77. R. Hoffmann, *J. Chem. Phys.*, **39**, 1397 (1963).
78. Reference 37, p. 102.
79. M. Galvàn, A. Dal Pino, Jr., J. Wang and J.D. Joannapoulos, *J. Phys. Chem.*, **97**, 783 (1993).
80. J. Tersoff and D.R. Hamann, *Phys. Rev.*, **1331**, 805 (1985).
81. B. Venkataraman, G.W. Flynn, J.L. Wilbur, J.P. Folkers and G.M. Whitesides, *J. Phys. Chem.*, **99**, 8684 (1995); D.M. Cyr, B. Venkataraman, G.W. Flynn, A. Black and G.M. Whitesides, ibid., **100**, 13747 (1996).
82. J.E. Chadwick, D.C. Myles and R.L. Garrett, *J. Am. Chem. Soc.*, **115**, 10364 (1993).
83. L.H. Lee, *New Trends in Physics and Physical Chemistry of Polymers*, Plenum, New York, 1989.
84. J. Harris and S. Anderson, *Phys. Rev. Lett.*, **55**, 1583 (1985).
85. M.L. Cohen, M.V. Ganduglia-Pirovano and J. Kudrnovsky, *Phys. Rev. Lett.*, **72**, 3222 (1994).
86. A. Corma, F. Llakis and C. Zicovich-Wilson, *J. Am. Chem. Soc.*, **116**, 134 (1994).
87. See Chapter 3 for a discussion of EEM.
88. R. Dronskowski, *Inorg. Chem.*, **31**, 3107 (1992); *J. Am. Chem. Soc.*, **113**, 6730 (1992).
89. A. Dal Pino, Jr., M. Galván, T.A. Arias and J.D. Joannopoulos, *J. Phys. Chem.*, **98**, 1606 (1993).
90. F. Mendizabal, R. Contreras and A. Aizman, *Int. J. Quantum Chem.*, **56**, 819 (1995).
91. J.E. Huheey, *J. Org. Chem.*, **36**, 204 (1971).
92. P. Politzer, *J. Chem. Phys.*, **86**, 1072 (1987).

6 Physical Hardness

INTRODUCTION

The physical hardness of a solid refers to its resistance to a change in shape or volume. Often the term "hard" is used as a synonym for "brittle", meaning susceptible to fracture. Soft solids are said to be malleable or ductile. Ionic and covalent solids are then hard and brittle. Metals are usually soft and malleable. The term "mechanical strength" is often used to describe hardness. The strength of materials is a subject of great practical importance.

Oddly enough, there has been no exact definition of hardness. Its value for a given sample is usually determined by very empirical methods, such as the scratch test, which gives the Moh scale of hardness, or the effects of dropping a weight on the sample. Such numbers are very useful, but difficult to interpret in a fundamental way. Also, the results are very dependent on the past history of the sample and its purity.

In the Vickers test for hardness, which is the most quantitative, the indentation left by a diamond stylus under a fixed load is measured. The hardness number can be expressed in pressure units, usually kg/mm^2. This test, and the scratch test, are irreversible. That is, the sample does not return to its original state. The deformation is said to be plastic, rather than elastic.

Mechanical strength is studied under the heading of elasticity.[1] This is the science of the response of a solid sample to applied forces. The forces are described by tensors, called stresses, which give the direction of the force and the crystal face to which it is applied. The responses, called strains, are also given by tensors which give the relative changes in dimensions or shape. The ratio of a stress to its corresponding strain is called an elastic modulus.

For small stresses the modulus is a constant and the material behaves elastically. It returns to its original condition when the stress is removed. For larger stresses, the elastic limit may be exceeded, and the sample undergoes a permanent, or plasic, deformation. Important stresses are compressional stresses, in which the force acts in one dimension only; hydrostatic, in which the force acts equally in all directions; and shearing, in which forces act to move parallel planes of the sample past each other.

At the microscopic level, shear stresses cause the gliding of planes of atoms over each other. This is the most common and easiest way for a solid to change its shape. The hardness, or the force needed, is very dependent on the presence of crystal defects. Even a pure crystal in the process of being formed will contain many defects, necessarily. The important defects for gliding motion are edge and screw dislocations. These are the front (back) and side (left/right) edges of a spreading region that is locally sheared.

A shearing stress will cause the dislocations to move, increasing the size of the locally sheared region. For the front edge the way in which this happens is illustrated in Figure 6.1. In projection, planes of atoms are shown as lines. Atom 2 in Figure 6.1 is at the center of the dislocation. By breaking the 3–6 bond and forming a 2–6 bond, the dislocation line has moved to the right. The actual movement occurs at "kinks" along the otherwise smooth line. Of course, it is accompanied by small changes in atomic positions.

Minute amounts of added material can change the strength greatly. For example, added carbon atoms in iron can act as dislocation traps and halt the gliding motion. Work-hardening of metals is a process whereby many of the dislocations intersect and collide with one another, thereby becoming partially immobilized. The movement of dislocations can be studied at various temperatures and the activation energy found. At 1500 °C plastic flow can be seen, even in diamond.

Gilman has called attention to the similarity of kink movement to a simple substitution reaction,[3]

$$\begin{array}{c} \mathrm{A{-}B + C = (A{-}B{-}C)^{\ddagger} \rightarrow A + B{-}C} \\ 3 - 6 + 2 = (3 - 6 - 2)^{\ddagger} \rightarrow 3 + 6 - 2 \end{array} \tag{6.1}$$

He argued that, just as in simple chemical reactions, the formation of the transition state, $(\mathrm{A{-}B{-}C})^{\ddagger}$ would lead to a raising of the energy of the HOMO and a lowerng of the energy of the LUMO. He then postulated that the gap between them vanishes at the transition state. The activation barrier for reaction (6.1) becomes $E_g^{\circ}/2$, as shown in Figure 6.2.

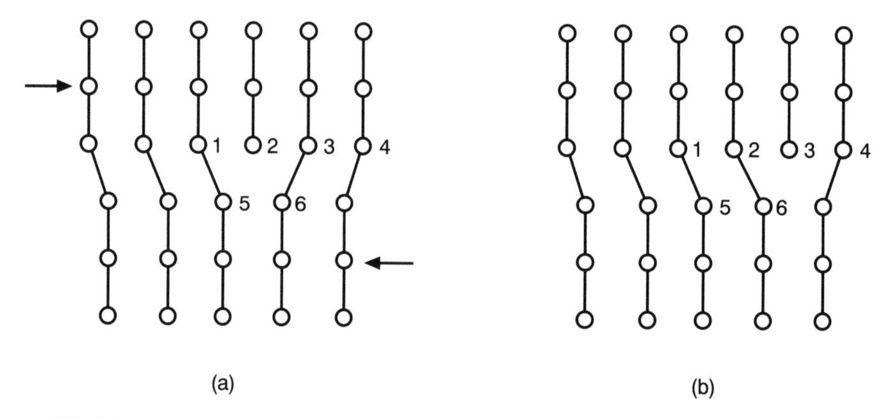

Figure 6.1 Movement of an edge dislocation, or kink, under the action of a shearing force (indicated by arrows). (a) Original bonding; (b) bonding after kink movement. After Reference 2.

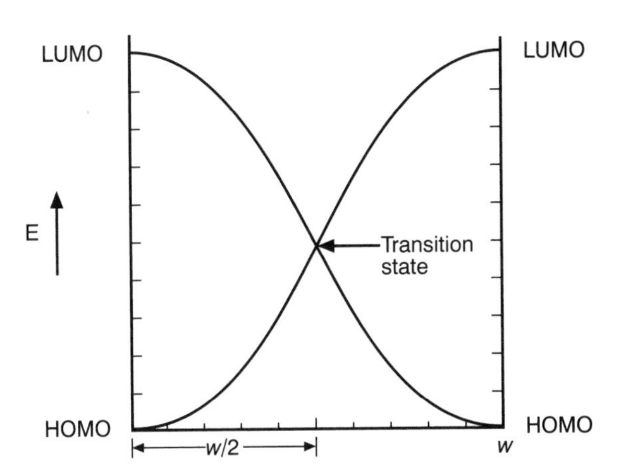

Figure 6.2 Orbital correlation diagram for kink movement, viewed as a simple substitution reaction. The width of the kink is w, and the reaction coordinate is in units of w. After Reference 3.

Taking the 4–4 covalent solids as examples, Figure 6.3 shows a graph of the glide activation energy plotted against the energy gap, E_g°, or the HOMO–LUMO gap. The correlation is remarkable and explains neatly why C and Si are hard and brittle, whereas α-tin is almost malleable. The slope is exactly 2.00, which is explained by realizing that the activation energy must include the energy of formation of new kinks. These are formed in pairs and then must move apart. The total energy needed is $4 \times E_g^\circ/2 = 2E_g^\circ$.[3]

Since we are taling about a solid with filled valence bands and empty conduction bands, the disappearance of the gap means that the kink system has become metallized by the shearing stress. Indeed, it is well known that many substances become metallic at high pressures.[4] When the atoms of a semiconductor, for example, are brought closer together, the orbital overlap increases, the band widths increase and the band gaps become smaller. If the band gaps become zero, the substance becomes a metal.

The usual criterion is that the molar volume, V, must become equal to the molar refractivity, $R = \frac{4}{3}\pi N_0 \alpha$, where α is the gas-phase polarizability. Then the Clausius–Mosotti equation becomes

$$\frac{\varepsilon - 1}{\varepsilon + 2} = \frac{R}{V} = 1, \qquad \varepsilon = \infty \tag{6.2}$$

The dielectric constant, ε, is infinite for a metal.

The transition pressure at which semiconductors become metals has been measured for a number of 4–4, 3–5 and 6–2 solids. Gilman showed that these numbers are a linear function of the hardness numbers measured by the Vickers test.[5] This test measures the resistance to a compressional force, acting in one

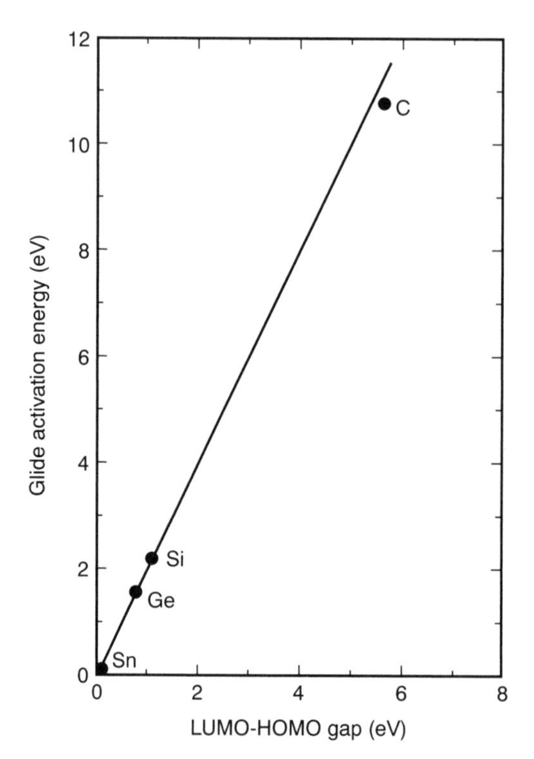

Figure 6.3 Plot of the glide activation energy against the band gap, E_g°, for the Group 14 elements. Reprinted with permission from J.J. Gilman, Science, *261*, 1436 (1993). © 1993 American Association for the Advancement of Science.

direction only. This leads to a rather different view of the metallization process. Instead of increased overlap being due to a shortening of bond lengths, it results from a change in bond angles.

How this happens is shown in Figure 6.4.[6] The semiconductors to which the above remarks apply all have the four-coordinate tetrahedral structures characteristic of covalent bonding. Applying a compressional force along the z-axis will cause a tetragonal distortion, as shown. There is a volume decrease of about 25 percent, but the nearest-neighbor distances are almost unchanged. Next-nearest neighbors are brought closer, however. Crystallographic data show that $\theta_0 = 109.5^\circ$, but that $\theta_1 = 149^\circ$ and $\theta_2 = 94^\circ$.

Although we could say that there is effectively more overlap because of an increased number of neighbors, there is an even more satisfying explanation. Consider again the change from the tetrahedral structure of CH_4 to a planar one, discussed at the beginning of Chapter 4. The energy of the HOMO increased and the energy of the LUMO decreased, as we went from tetrahedral to planar. The same changes in a tetrahedral solid, as shown in Figure 6.4, will cause E_g° to decrease. Simple Hückel theory leads to a prediction that θ_1 should equal 180° and

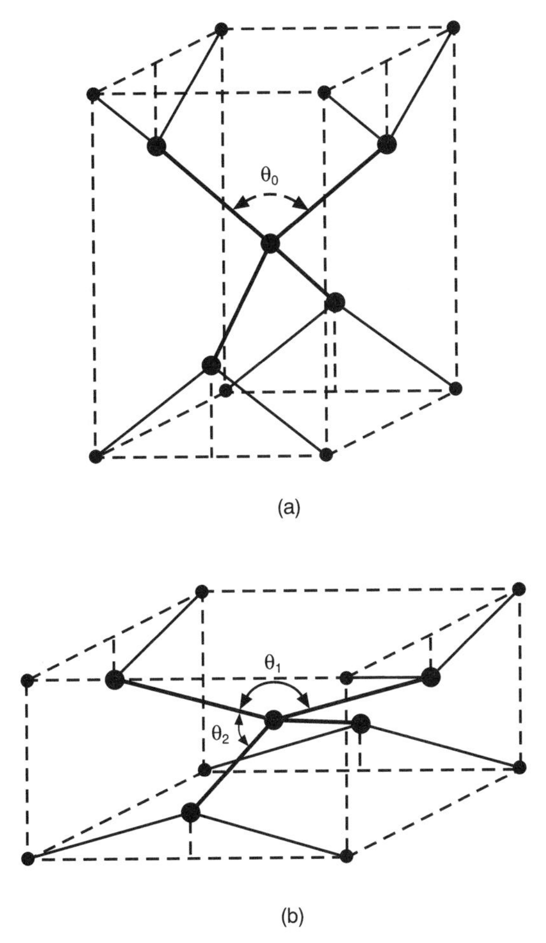

Figure 6.4 Schematic drawing of the change from (a) the diamond structure of α-tin to (b) the structure of metallic β-tin. For clarity, five of the atoms in the unit cell have been drawn larger than the others. After Reference 6.

θ_2 should be 90° for the gap to become zero. But this does not include the effects of the band width, which will make the required changes smaller, as observed.

The energies required to change bond angles and bond lengths in solids can be estimated from force constants for these changes. The force constants come from measurement of the elastic constants.[7] As expected, it is easier to change bond angles than bond lengths. Therefore metallization by bond-angle changes is easier than by bond-length changes. The stresses needed, called shearing stresses, are less than compressional (dilational) stresses. A portion of a solid that is metallic is a region where local bonding has vanished, or can fluctuate readily. Chemically, it is a highly reactive zone.

This leads to a number of interesing possibilities for new applications. One is the interpretation of the detonations of solid explosives.[8] These usually decompose by the very rapid propagation of a detonation front. The reactions are too fast to be thermally activated. Instead the high pressure in a front can cause local metallization, allowing rapid chemical reaction. This can occur if Equation (6.2) is satisfied, or if the HOMO–LUMO gap becomes zero by bond-bending.

A number of explosives fit the latter mechanism. These include lead azide, ammonium nitrate and pentaerythritol nitrate.[8,9] The critical strains needed to initiate explosion may be estimated. As expected, the sensitivity of explosives to shock can also be rationalized by estimating their initial HOMO–LUMO gaps, or E_g°. For solids such as $Pb(N_3)_2$, the energy needed to excite an electron from the valence band to the conduction band can be found by spectroscopy. A small gap means high sensitivity.[10]

For organic solids the individual molecular properties are used. For explosives containing aromatic rings, there is a correlation between the sensitivity and the resonance energy.[11] The smaller the resonance energy per electron (REPE), the more sensitive is the compound. All these examples show the expected relationship: the softer the solid (in chemical terms), the more reactive it is.

There are many other examples of chemical reactions being induced by shearing stresses.[9] A mechanism involving metallization seems plausible. Areas of application include photochemistry, degradation of polymers, friction and wear, mechanical alloying and cutting processes.

A DEFINITION OF PHYSICAL HARDNESS

Each of the elastic constants is a measure of hardness for a particular deformation. A modulus measures the force, or pressure, needed to cause a certain change in shape or volume. But there is one modulus in particular that seems best suited to be a general measure of hardness. This is the bulk modulus, B, which determines the volume change for a sample under hydrostatic pressure. It is defined as

$$B = -V\left(\frac{\partial P}{\partial V}\right)_T = \frac{1}{\kappa} \tag{6.3}$$

The reciprocal of the bulk modulus is the compressibility, κ.

For an isotropic solid, a force exerted equally in three dimensions will produce a change in volume by a uniform decrease of distances between nearest neighbors. Bond angles will not be changed. The inverse relationship between B and κ resembles the relationship between hardness and softness. Certainly one expects a solid which is physically soft to be compressible, and one which is hard to resist compression.

However, there is an even better reason to single out B as a hardness factor. In classical thermodynamics there is a standard equation[12]

$$\left(\frac{\partial \mu}{\partial N}\right)_{T,V} = \frac{V}{N^2 \kappa} = BV_0 \tag{6.4}$$

where V_0 is the molar volume and N is the number of molecules. The factor of N^2 is needed for dimensional purposes. B has the units of pressure and BV_0 has the units of energy. If V_0 is calculated per mole of atoms, then N^2 may be ignored.

Because of Equation (6.4) Yang et al. have proposed that BV_0 be called the physical hardness, H.[13] It has the units of energy, the same as chemical hardness. Its reciprocal is the softness, proportional to κ, as desired. Their proposal was strongly reinforced by showing that BV_0 for a number of substances followed much the same ordering as the Moh hardness of those substances.

Table 6.1 contains experimental values of $H = BV_0$ for solids of simple type, with cubic structures. For comparison, the values of the cohesive energy, ΔE_{coh}, are also included. The hardness numbers are of the same magnitude as the energies of atomization. There is also a definite trend for H to increase as ΔE_{coh} increases. However it is not very regular. For ionic solids, H is almost always less than the cohesive energy. For covalent solids, including metals, H is always much greater than ΔE_{coh}, except for Li.

We also see that W and Pt, and a few other noble metals, have H greater than carbon. On this scale they are harder than diamond, which means that H no longer matches the scratch test for hardness. This does not invalidate BV_0 as a legitimate scale for hardness. It is well defined, has a thermodynamic basis, and measures the resistance to well-defined changes. But for very hard substances, it is not equivalent to the Moh scale, which measures plastic hardness. Actually B, rather than BV_0, matches the scratch test best.[14]

Though not apparent from Table 6.1, there is a relationship between B and ΔE_{coh}.[15]

$$B = C\left(\frac{\Delta E_{\text{coh}} m}{Z V_0 q}\right)^X \tag{6.5}$$

where C is a constant, m is the number of component elements, q is the number of atoms per molecule and Z is the maximum valence, or the number of bonds per molecule. The exponent, X is about 2.9, and the value of C depends on the units chosen. Equation (6.5) fits both ionic and covalent solids, but not metals. It is strictly empirical, and its origin is unclear.

To undertand better the way in which H and ΔE_{coh} are connected, we turn again to the simple theory of bonding developed in **Chapter 4**.

Table 1 Hardness Numbers, BV_0, for Simple Solids[a]

Element	H [kcal/mol]	ΔE_{coh} [kcal/mol]	Element	H [kcal/mol]	ΔE_{coh} [kcal/mol]
Li	38	38	Zn	128	31
Na	39	26	Ag	246	68
Cs	35	19	Cd	137	27
Mg	113	35	In	147	58
Al	173	78	Pb	180	47
Ca	107	43	W	790	202
Si	100	39	Pt	651	135
Cr	332	95	C	478	171
Mn	250	67	Si	282	109
Fe	280	99	Ge	270	90
Cu	235	81	Sn(β)	285	73
Compound	H [kcal/mol]	ΔE_{coh} [kcal/mol]	**Compound**	H [kcal/mol]	ΔE_{coh} [kcal/mol]
LiF	78	204	BaS	165	218
NaCl	79	153	ZnO	249	174
KCl	81	155	ZnS	217	147
KI	53	124	ZnTe	186	107
RbF	48	171	CdTe	208	97
CuBr	119	133	InSb	218	128
AgCl	136	127	InP	264	155
MgO	207	239	GaAs	246	156
CaO	229	254	GaSb	230	139
SrO	217	240	SiC	367	269
MnO	240	219	TiC	349	328

[a] Data from References 13 and 21. ΔE_{coh} for compounds should be divided by two to compare with elements.

For ionic bonding we have

$$U = -\frac{MZ^2}{R} + mB\,e^{-R/\rho} \tag{6.6}$$

$$U_0 = -\frac{MZ^2}{R_0}(1 - \rho/R_0) \tag{6.7}$$

$$BV_0 = \frac{MZ^2}{9R_0}(-2 + R_0/\rho) \tag{6.8}$$

For covalent bonding

$$U = -2C\,\mathrm{e}^{-R/2\rho} + mB\,\mathrm{e}^{-R/\rho} \tag{6.9}$$

$$U_0 = -C\,\mathrm{e}^{-R_0/2\rho} \tag{6.10a}$$

$$BV_0 = U_0 R_0^2/18\rho^2 \tag{6.10b}$$

Equations (6.8) and (6.10(b)) show that there is a connection between H and the cohesive energy. Unfortunately we cannot readily calculate H from ΔE_{coh} because we do not know ρ, except as calculated from Equations (6.8) and (6.10(b)). The values of ρ found in this way increase as R_0 increases, and are about five to seven times smaller than R_0. The cohesive energy for covalent bonding is equal to U_0, but for ionic bonding we must correct U_0 by the energy released when the separated ions change to atoms.

Equations (6.8) and (6.10(b)) come from the definition of B in Equation (6.3), and the relations

$$V = cR^3; \qquad \mathrm{d}V/V = 3\mathrm{d}R/R; \qquad P = -(\partial U/\partial V)_T \tag{6.11}$$

where c is an inconsequential constant. For cubic crystals we can relate changes in the volume to changes in the nearest-neighbor distance, R, in a simple way. We then get

$$B = -V(\partial P/\partial V)_T = \frac{V^2}{V}(\partial^2 U/\partial V^2)_T = \frac{R_0^2}{9V_0}(\partial^2 U/\partial R^2)_T \tag{6.12}$$

The curvature $(\partial^2 U/\partial R^2)_T$ may be identified with the force constant, f, when U is the potential energy function for a diatomic molecule. It has also been identified as the single most important property determining the scratch hardness of a solid.[14] The relationship between the hardness, H, and the cohesive energy now seems to be the same as that between the force constant of a diatomic molecule and its dissociation energy, D_0. When one increases, the other usually does also. But there is no simple relation between them.[16] Force constants depend more strongly on the repulsive part of the potential energy function than bond energies do.

Equation (6.12) suggests that we try to calculate BV_0 from the point of view of force constants for the bonds undergoing compression. This has already been done for cubic crystals by Pauling and Waser.[17] Their procedure was to equate the pressure–volume work done to the energy of compressing the bonds. Their result was

$$BV_0 = \frac{nfR_0^2}{9} \tag{6.13}$$

Table 6.2 Force Constants for Solids and Corresponding Diatomic Molecules

	$10^5 f$		f
Li(s)	0.059	KI(s)	0.124
Li_2(g)	0.255	KI(g)	0.610
CuCl(s)	0.65	C(s)	6.09
CuCl(g)	2.31	C_2(g)	12.20

f is in millidynes/cm.

where $n = m/2$, the number of bonds per atom. The coordnation number m is divided by two because each bond is shared by two atoms.

Equation (6.13) is completely consistent with Equation (6.12), since the total change in energy depends on the number of bonds, as well as their force constants. V_0 is the volume per mole of atoms, as before. The force constants for a number of solids have been calculated from Equation (6.13).[17] Comparing them with the force constants for the related diatomic molecules gives the results in Table 6.2.

The force constant for a solid is only a fraction of that for the molecule. The reason, of course, is that the average bond energy for the solid is only a fraction of that for the diatomic case. For metals there are too many bonds and not enough electrons. For ionic solids, the charges tend to cancel each other. Only for the covalent 4–4 solids, such as carbon (diamond) and silicon, are the bonds in the solid equivalent to those in small molecules. Actually C_2 has a double bond. The force constant for a single C–C bond is about 5 mdyn/cm, which is to be compared with f for the solid. The 3–5, 2–6 and 1–7 covalent cases give reduced bonding compared with the 4–4, as discussed in Chapter 5.

THE PRINCIPLE OF MAXIMUM PHYSICAL HARDNESS, PMPH

A great advanage of using the thermodynamic definition of hardness is that we can find other thermodynamic equations involving H. For example, taking a grand canonical ensemble, there is a well-known equation for the fluctuations in the number of particles, N,[12]

$$(\partial N/\partial \mu)_{V,T} = \frac{\kappa N^2}{V} = \beta \langle (N - \langle N \rangle)^2 \rangle \tag{6.14}$$

where $\beta = 1/kT$. Since $\kappa/V_0 = 1/H$, we can say that crystals which are physically soft have large fluctuations in N. In this case the systems of the ensemble are crytals of identical volume, but with varying numbers of component atoms. The average value $\langle N \rangle$ is a constant, N°.

Using the proof of Chattaraj and Parr, as was done in Chapter 4, we can conclude that the mechanical softness is a minimum, and the hardness is a maximum, for the equilibrium state. This would apply to any condensed system, solid or liquid. This is a very reasonable result. Chemical hardness is the resistance to change in the electron distribution, and physical hardness measures the resistance to change of the nuclear positions. An equilibrium system should have the greatest resistance to change for both of these properties.

To test this prediction for solids, we must show that BV_0 is a maximum for the equilibrium state of the crystal. That is, $\delta(BV_0) = 0$, subject to certain restrictions, such as constant volume and temperature. Another restriction is made: to consider only cubic crystals containing a single element, or the binary AB compounds. This is so that we can use the information in Equations (6.11) and (6.12). We wish to prove that the equilibrium crystal has $V = V_0$, the experimental volume per mole of atoms.

Differentiate BV with respect to V, using the definition of B in Equation (6.3) and the definition of P in Equation (6.11).

$$B + V(\partial B/\partial V)_{T,V_0} = 0; \qquad -2(\partial^2 U/\partial V^2) = V_0(\partial^3 U/\partial V^3) \tag{6.15}$$

With the relation between V and R in Equation (6.11) this can be written as

$$\frac{-2R_0^2}{9}\left(\frac{\partial^2 U}{\partial R^2}\right)_{T,R_0} = \frac{R_0^3}{27}\left(\frac{\partial^3 U}{\partial R^3}\right)_{T,R_0} \tag{6.16}$$

Although Equation (6.6) and (6.9) give a functional dependence of U on R, they are not accurate enough to give the third derivative. Instead U is expanded as a power series in $x = (R - R_0)$, where x is small:

$$U = U_0 + \frac{f}{2}x^2 + gx^3 + hx^4 \cdots \tag{6.17}$$

The parameter f is the force constant and g is the anharmonicity constant. We have no term linear in x at equilibrium. That is $(\partial U/\partial R) = 0$, $(\partial^2 U/\partial R^2) = f$ and $(\partial^3 U/\partial R^3) = 6g$ at $R = R_0$. Putting this information into Equation (6.16) gives us $f = -gR_0$. The PMPH gives us a defnite relationship between the force constant and the anharmonicity constant. We know that, since f is always positive, the negative sign for g is correct, from experimental evidence to be considered next.

There are two properties of solids which depend on the anharmonicity term most directly. One is the coefficient of thermal expansion, and the other is the variation of the bulk modulus with pressure, $(\partial B/\partial P)_T$. The experimental results in the latter case are given either by a power-series equation

$$V = V_0(1 - aP + bP^2) \tag{6.18}$$

or by the Murnaghan equation:[19]

$$B_1 \ln\left(\frac{V}{V_0}\right) = -\ln\left(1 + \frac{B_1 P}{B_0}\right) \tag{6.19}$$

Expanding the logs in Equation (6.19) for moderate pressures, we obtain Equation (6.18) again, with $a = 1/B_0$ and $b = (1 + B_1)/2B_0^2$. B_0 is the modulus at zero pressure and B_1 is $(\partial B/\partial P)$.

Using Equation (6.18) we can solve most easily for the compressibility, κ, and its pressure variation:

$$\kappa = a - 2bP + a^2 P \tag{6.20}$$

$$\left(\frac{\partial \kappa}{\partial P}\right) = -2b + a^2 = -\kappa^2\left(\frac{\partial B}{\partial P}\right) \tag{6.21}$$

Finally we solve for $(\partial B/\partial P)$, at moderate pressure:

$$\left(\frac{\partial B}{\partial P}\right) = \left(\frac{2b}{a^2} - 1\right) = B_1 \tag{6.22}$$

Experiments show that the modulus is indeed a linear function of the pressure, over the range of pressures available. The slope is given by Equation (6.22). The intercept is B_0.

Pauling and Waser interpreted both B_0 and B_1 in terms of force constants.[17] In Equation (6.13) it is B_0 that appears and gives a value for f. B_1 then gives the value of g, again equating pressure–volume work to the energy of compressing the bonds. They found that[20]

$$g = (1 - b/a^2)f/2R_0 \tag{6.23}$$

Using the result from the PMPH, $g = -f/R_0$, we find $(b/a^2) = 3$ and $B_1 = 5.0$. Therefore, for cubic solids, the pressure derivative of the modulus is equal to a constant and dimensionless number, 5.0.

The value of a, the compressibility at very low pressure, is well known for many solids. However, b is difficult to determine. Older, static methods are not reliable. Better results are obtained by ultrasonic pulse methods on single crystals, but even here different investigators can differ by 10–20 percent. Table 6.3 shows

Table 6.3 Pressure Derivative of the Bulk Modulus for Single Crystals at Room Temperature[a]

Substance	$\partial B/\partial P$	Substance	$\partial B/\partial P$
Al	4.4	MgO	4.4
BaF_2	5.1	KBr	5.4
CaF_2	5.5	KCl	5.4
CsBr	5.1	KF	5.0
CsI	5.4	RbBr	5.3[b]
Cu	5.5	RbCl	5.5
GaSb	4.7	RbI	5.3
GaAs	4.5	LiF	4.8
Au	5.2	NaBr	5.0
Fe	5.3	NaCl	5.1
Pb	5.5	NaF	5.2
Ag	5.1		

[a] See Reference 18 for sources of data.
[b] At 220 A.

experimental values of $(\partial B/\partial P)$ at room temperature.[21] They have been averaged when several values are available and, in one or two cases, values markedly different from the mean have been dropped.

The closeness of these numbers to the predicted value of 5.00 is indeed remarkable. It may well be that the theoretical value is more reliable than some of the experimental values. However, there are approximations in the theory. For example, writing the pressure as $(\partial U/\partial V)_T$ ignores the contributions of the lattice vibrations to the pressure. This term should be small, but it will contribute varying amounts for different substances.

If the potential energy function of a solid were simply harmonic, there would be no thermal expansion:

$$\alpha = \frac{1}{3V}\left(\frac{\partial V}{\partial R}\right)_P = \frac{1}{R_0}\left(\frac{\partial \bar{x}}{\partial T}\right) = 0 \tag{6.24}$$

The average displacement, $\bar{x} = \overline{(R - R_0)}$, would be zero. But the term in x^3, Equation (6.17), makes a positive value of x more probable than a negative value. This is based on the Boltzmann distribution law, and a negative value for g. A calculation gives

$$\bar{x} = \frac{3kT|g|}{f^2}; \qquad \alpha = \frac{C_v|g|}{f^2 R_0} = \frac{C_v}{fR_0^2} \tag{6.25}$$

In the last equality, we have used the maximum hardness result, $-f = gR_0$.

We can use Equation (6.13) to eliminate fR_0^2, and relate α, the coefficient of thermal expansion, to κ, the compressibility. The value of α could then be calculated. But experimentally it is better to use the ratio of α to κ, by introducing the Grüneisen constant, γ[22]:

$$\gamma = \frac{3\alpha V_0}{C_V \kappa} \tag{6.26}$$

Theoretically γ was first defined as the variation of the eigenfrequencies of a solid as the volume changed. Equation (6.26) can be derived from this definition, using the model of an elastic continuum. The value of γ is not predicted, but operationally it can be measured using Equation (6.26). For many solids C_V is not equal to $3k$, as in Equation (6.25), because the highest frequencies are not excited. This causes no error, since $\bar{x}$ is not increased by such frequencies either.

Combining Equations (6.13), (6.25) and (6.26), we can find the value of γ. The result is a simple and novel one, $\gamma = n/3$. Thus we predict that $\gamma = 0.667$ for crystals with the zinc-blende or wurtzite structure, $\gamma = 2.00$ for cubic close packing, and so on. Table 6.4 contains experimental values of the Grüneisen constant for some covalent solids. The number can be compared with the predicted $n/3$. Although the agreement is not quantitative, the trend with changing coordination number is unmistakable. For example, the bcc metals have an average value of $\gamma = 1.46$, consistent with $n = 4$. The fcc metals have an average value of $\gamma = 2.17$, consistent with $n = 6$.

Table 6.4 Grüneisen Constants for Some Covalent Solids at Room Temperature[a]

Substance	γ	$n/3$	Substance	γ	$n/3$
C	0.86	0.67	Co	1.87	2.00
Si	0.45	0.67	Ni	1.88	2.00
Ge	0.72	0.67	Cu	1.86	2.00
Li	1.17	1.33	Pd	2.23	2.00
Na	1.25	1.33	Ag	2.40	2.00
K	1.34	1.33	Pt	2.54	2.00
Rb	1.48	1.33	Au	2.40	2.00
Ca	1.29	1.33	Pb	2.23	2.00
W	1.62	1.33	CdS	0.49	0.67
Zn	1.57	1.33	CuCl	0.84	0.67
Mo	1.57	1.33	GaAs	0.72	0.67
Ta	1.75	1.33	InAs	0.54	0.67
Fe	1.60	1.33	InSb	0.50	0.67
Al	2.17	2.00	ZnO	0.66	0.67
			ZnS	0.76	0.67

[a] See Reference 18 for data sources.

Equation (6.13) was derived on a model of covalent bonds between nearest neighbors. It is not strictly applicable to ionic solids. The repulsion part of the potential energy must be similar for ionic and covalent cases, but the attraction part for ionic solids must also include the sum of the coulombic interactions with the remainder of the lattice. In effect, the number of bonds is increased. To see the magnitude of this effort, compare Equations (6.7) and (6.8) with their counterparts for a diatomic molecule, or ion-pair.

The latter is less by a factor of M, the Madelung constant. Then $(M-1)$ must be the effect of the rest of the lattice. This suggests the "corrected" value $\gamma = nM/3$ for ionic solids.

Table 6.5 shows the "corrected" results for some ionic solids. The assumption is made that solids with coordination number 6 (rock-salt structure) and coordination number 8 (CsCl structure) are sufficiently ionic for the "corrected" values to be needed. This expectation is clearly met. The larger values of γ for CN8 compared with CN6 are found, as expected.

Only BeO has CN4, yet it has a value of γ that shows ionic bonding. This agrees with the calculation of the cohesive energy presented ealier. CN4 is forced upon BeO by the radius ratio effect. The value of γ forBeO may be compared with the γ for ZnO, which is just what is expected for covalent bonding.

The theory so far has only shown that BV_0 has an extremum value, not that it is a maximum. The answer to that question lies in the value of h in Equation (1.17). This number must be negative, or have a small positive value, in order for BV_0 to be a maximum. There is very little evidence concerning the magnitude, or even the sign, of the quartic term. It is usually considred to be a "softening" term (h negative).[22] This is supported by some heat capacity data at high temperatures. The Born–Mayer and Morse equations are not accurate enough to predict g, much less h.

At high pressures, BV is usually larger than BV_0 at atmospheric pressure. But this is not a contradiction. The essential reason for a maximum in BV_0 is that the energy is a minimum at R_0, so long as the crystal symmetry is maintained. At higher pressures, values of R different from R_0 will be the new equilibrium ones. In all cases, values of the distances different from the equilibrium ones will lead to an increase in energy and a decrease in BV. Constant pressure is a constraint on the PMPH.

If one attempts to show that BV_0 is a maximum for the liquid state, the results are poor.[23] Reasonable values of the coordination number cannot be found from experimental values of γ. Also, $(\partial B/\partial P)$ for liquids is about 10 for liquids, and not 5.[24]

There are two main reasons for this failure. One is that the thermal pressure can no longer be ignored, compared with the pressure due to lattice expansion. That is, thermal energies are comparable with the intermolecular energies. The second reason is that volume changes in liquids are largely due to changes in the number of "holes", and not due to uniform changes in R_0. Unfortunately, the theory of liquids is too complicated to correct easily for these factors.

Table 6.5 Grüneisen Constants for Some Ionic Solids at Room Temperature[a]

Substance	γ	$nM/3$	Substance	γ	$nM/3$
BeO	1.54	1.10	AgCl	1.90	1.75
LiF	1.58	1.75	AgBr	2.05	1.75
LiCl	1.54	1.75	MgO	1.59	1.75
NaF	1.51	1.75	CaO	1.51	1.75
NaCl	1.57	1.75	SrO	1.52	1.75
NaBr	1.57	1.75	CsF	1.49	1.75
NaI	1.71	1.75	CsBr	1.93	2.34
KF	1.48	1.75	CsCl	1.97	2.34
KCl	1.45	1.75	CsI	2.00	2.34
KBr	1.43	1.75	TlCl	2.30	2.34
KI	1.47	1.75	TlBr	2.19	2.34
RbI	1.50	1.75			

[a] Data from Reference 18.

There seems to be a law of nature that, in an equilibrium system, the chemical hardness and the physical hardness have maximum values, compared with nearby non-equilibrium states. However, it must not be inferred that these maximum principles are being proposed to take the place of estabished criteria for equilibrium. Instead, they are necessary consequences of these fundamental laws. It is very clear that the Principle of Maximum Hardness for electrons is a result of the quantum mechanical criterion of minimum energy. Similarly, Sanchez has recently derived the relationship $(\partial B/\partial P) = 5$ by a straightforward manipulation of the thermodynamic equation of state.[25] The PMPH is a result of the laws of thermodynamics.

It is, of course, very reasonable that the equilibrium state has the greatest resistance to change, both of electron distribution and of nuclear positions. After all, the non-equilibrium states must all change these properties to reach equilibrium. DFT has again provided new insights into chemical behavior. It should be highly worthwhile to apply DFT to the coupled variations of the nuclear posiions and ρ, the density function. There are already activities in this area that are yielding interesting results.

One approach has been to define nuclear softness functions, σ_α, and nuclear reactivity indices, f_α:[26]

$$\sigma_\alpha = \left(\frac{\partial F_\alpha}{\partial \mu}\right)_v; \qquad f_\alpha = \left(\frac{\partial F_\alpha}{\partial N}\right)_v \tag{6.27}$$

F_α is the force exerted on nucleus α by the electron cloud. The reactivity index is also a nuclear Fukui function. Although it is completely general, Equation (6.27) has been applied mainly to surface atoms.

Another approach has been to develop mapping procedures connecting changes in ρ with changes in nuclear positions.[27] The latter are selected as the normal modes of vibration which are bond-stretching. The hardness of the local electron cloud increases with increasing force constant. As might be expected, there is a close connection between the physical hardness (nuclear motion) and the chemical hardness.

A third approach has been to identify a nuclear reactivity index, $h(r)$, which defines a local hardness:[28]

$$h(r) = \left(\frac{\delta\mu}{\delta\rho}\right)_{N,T} = \beta[\langle\mu \cdot V - \langle\mu\rangle\langle v\rangle\rangle] \tag{6.28}$$

A large local value of $h(r)$ means a large resistance to change of the shape of the local electron density. It involves fluctuations in nuclear positions, e.g. in vibrations. The averaging in Equation (6.28) is over an ensemble. Since the covariance of μ and v does not necessarily have a fixed sign, we cannot say that $h(r)$ is a maximum or minimum.

Equilibrium for a solid would also stipulate the shape, since the surface free energy should be a minimum. For an isotropic crystal this shape would be that of a sphere, but this is hardly ever a factor. The loss in energy for a surface atom is about one-half of the cohesive energy per atom. But the number of atoms on a typical surface is only about 10^{16}, or 10^{-8} mol, so the surface energy is very small. There is one observable effect, however; a collection of small crystals will cohere to form larger crystals, if a mechanism, such as digestion of a precipitate, is provided.

THE HARDNESS OF MOLECULES

The values of H listed in Table 6.1 for the physical hardness of solids raise an interesting question. Should there not be a corresponding number, H', for the physical hardness of molecules? After all, there are force constants in molecules as well as in solids. Equation (6.13) might serve for a diatomic molecule, if n were simply set equal to one.

There are many difficulties with this approach. The volume of a single molecule is not a well-defined property, nor is the way in which it changes as R_0 changes. Also, we can hardly expect the fluctuation formula, Equation (6.14) to be valid, since we cannot change the number of molecules in an ensemble whose systems are single molecules. Therefore our proof of the PMH is not applicable.

We can still draw an interesting conclusion, if we assume that V and $(\partial V/\partial R)$ for a molecule are unknown but definite quantities, and that the Principle of

Maximum Physical Hardness is obeyed. Setting $(\partial V/\partial R) = V'$, and using Equation (6.15), we find

$$BV = -\left(\frac{V}{V'}\right)^2 \left(\frac{\partial^2 U}{\partial R^2}\right); \qquad 2\left(\frac{V}{V'}\right)^2 \left(\frac{\partial^2 U}{\partial R^2}\right) = -\left(\frac{V}{V'}\right)^3 \left(\frac{\partial^3 U}{\partial R^3}\right) \tag{6.29}$$

If we use Equation (6.17), and if the PMPH is valud, then $(V/V') = -f/3g$ and we can write

$$BV_0 = H' = f^3/9g^2 \tag{6.30}$$

Note that this definition is also true for solids, if $f = -gR_0$.

Equation (6.30) is a reasonable result for molecular hardness. Certainly H' should increase as the force constant increases, and decrease as $|g|$ increases. Remember g, being negative, is a softening factor. While we cannot test Equation (6.30) in the same way as for solids, we can calculate H' for a number of diatomic molecules, to see if they are in some way informative.

In place of equation (6.17), the potential energy of a diatomic molecule is usually given in the Dunham formulation:[29]

$$U = a_0\left(\frac{x}{R_0}\right)^2 \left[1 + \left(\frac{x}{R_0}\right)a_1 + \left(\frac{x}{R_0}\right)^2 a_2 \cdots\right] \tag{6.31}$$

The hardness can be written as $H' = 4fR_0^2/9a_1^2$. The values of a_0, a_1 and a_2 can be found from a detailed analysis of the rotation–vibration spectra of the molecule. The experimental results are presented as ω_e, the vibrational frequency, B_e, the rotational constant, $\omega_e X_e$, the anharmonicity constant, and α_e the rotation–vibration coupling constant. The subscript e refers to the ground-state or equilibrium value.[30]

Formulas are available to convert these constants into a_0, a_1 and a_2.[31] We have $a_0 = fR_0^2/2$, $a_1 = 2gR_0/f$, and $a_2 = 2hR_0^2/f$. The necessary constants have been obtained for a very large number of diatomic gas-phase molecules, both stable and unstable.[32] Table 6.6 gives some typical results for H'.

Table 6.6 gives the values of D_0, the gas-phase dissociation energies. It can be seen that H' is very similar to D_0: the correlation is quite good, though far from perfect. Ionic molecules have D_0 values larger than H', and covalent molecules have D_0 less than H', as a rule. Comparison of BeO and BeS with CaO and CaS suggests that the former are much more covalent than the latter.

We may also compare the relationships between D_0 and H', between D_0 and f, and between D_0 and η. Whereas there is a rough correlation in the latter cases, that between the dissociation energy and the physical hardness is the best, at least for covalent molecules. For example, F_2 and I_2 have nearly the same dissociation energies, but the force constant for F_2 is 2.75 times as large as that for I_2. Also, η for F_2 is 7.3 eV and that for I_2 is 3.8 eV, taking the vertical values.

Table 6.6 Hardness Numbers, H' for Some Diatomic Molecules[a]

	H' [kcal/mol]	D_0 [kcal/mol]		H' [kcal/mol]	D_0 [kcal/mol]
Ionic molecules					
LiF	54	136	CsI	33	82
LiI	48	82	BeO	114	106
LiH	46	56	BeS	104	88
NaCl	41	98	CaO	81	110
KBr	35	90	CaS	82	80
Covalent molecules					
H_2^+	38	61	B_2	50	70
H_2	79	103	C_2	141	143
BeH	66	54	N_2	243	225
BH	65	79	O_2	122	118
CH	74	81	F_2	45	37
NH	79	86	Cl_2	73	57
OH	91	102	I_2	47	36
HF	103	136	Na_2	25	17
HI	78	70	HF^+	56	79

[a] Data from Reference 32.

Compared with solids, we now have accurate values for the constant h in Equation (6.17). The range of a_1 and a_2 data is demonstrated by the examples in Table 6.7. In the case of solids, a_1 would have the constant value of -2.00.

Since we know h, we can at least test to see if BV_0 is a maximum or a minimum. If $(\partial H'/\partial V) = 0$, then $(\partial^2 H'/\partial V^2) < 0$, if we are at a maximum of H' at $V = V_0$. This works out to require that

$$\tfrac{9}{4}g^2 > fh; \qquad \tfrac{9}{8}a_1^2 > a_2 \tag{6.32}$$

This is true in all the examples in Table 6.7, and appears to be true in general. However, it should be borne in mind that there is no direct proof for the assumption of maximum physical hardness for molecules.

Nevertheless, it seems very natural that, just as for solids, molecules should have maximum chemical hardness and maximum physical hardness, at equilibrium. Polyatomic molecules offer a more difficult problem, since there are different force constants for the different normal modes. Also there are different anharmonicity constants, which need not be negative.

A natural choice for the physical hardness of a moelcule would be the totally symmetric breathing mode. This would correspond to the selection of the bulk modulus for solids. In Chapter 4 it was shown that the Principle of Maximum Chemical Hardness applied to antisymmetric vibrational modes of a molecule, but not to the symmetric modes. It would be of interest to see whether the Principle of Maximum Physical Hardness governed the symmetric vibrations of molecules.

Table 6.7 Values of $-a_1$ and a_2 for Some Gas-Phase Molecules

Molecule	$-a_1$	a_2	Molecule	$-a_1$	a_2
H_2	1.60	1.87	BeS	2.77	5.26
LiH	1.89	2.41	O_2	3.00	5.72
Cs_2	2.14	1.42	NaCl	3.08	6.51
HF	2.25	3.47	CaO	3.08	4.67
HI	2.57	4.16	Br_2	3.57	7.14
N_2	2.70	4.36	I_2	4.10	15.43

A relationship between the bonding in diatomic molecules and in solids has been demonstrated.[33] It follows from the so-called Universal Binding Energy Relation (UBER), applicable to both metals and covalent diatomic molecules. The energy and the interatomic separation are scaled in the following way:

$$E^*(R) = E(R)/\Delta E_{\mathrm{coh}} \tag{6.33}$$

$$R^* = (R - R_0)/l \tag{6.34}$$

The quantity l is the scaling length. One way to define l is to set $fl^2 = \Delta E_{\mathrm{coh}}$.

The remarkable result is that a plot of E^* vs. R^* is a universal one, in that a single curve fits the data for many metals and covalent molecules. It is also valid for data on adhesion, adsorption and impurity binding energies. It can be used to correlate a number of physical properties of metals, such as surface energies and equations of state. Note that cohesive energies and compressibility data are needed as input parameters. Also, UBER does not apply to ionic solids, nor to polar molecules.[34]

The first five chapters of this volume introduced the subject of chemical hardness, that is, the resistance to changes in the electron density function of a chemical system. The nuclei were supposedly held fixed in position. In spite of this limitation, a number of applications of chemical hardness to a better understanding of bonding energies, rates of reaction and structures, were given.

The fifth chapter, on solids, led to a consideraton of physical hardness, which has been covered in some detail in this chapter. A scale has been proposed, which is not the same as those used by material scientists, but which seems better related to chemistry. Since both chemical reactions and physical hardness require changes in nuclear positions, there should be a relationship between the two. For an engineer, changes in shape are more important than changes in volume.

It would be useful to have a measure of physical hardness for individual molecules. The foregoing is an attempt to provide such a measure. In the long run, it may be that some other approach will prove more useful. For example, it could be that starting with the chemical hardness, and modifying it for changes in nuclear positions, will give a general function for molecules. Such a function should be related to both the physical and chemical stability of chemical systems.

REFERENCES

1. For an introduction, see any elementary physics textbook. For more detail relevant to the present work see J.H. Wernick, *Treatise on Solid State Chemistry*, N.B. Hannay, Ed., Plenum Press, New York, 1975, Vol. 1, Chapter 4; E. Nembach, ibid., Vol. 2, Chapter 7.
2. For a good discussion of defects and their motion, see A.R West, *Basic Solid State Chemistry*, John Wiley, New York, 1984.
3. J.J. Gilman, *Science*, **261**, 1436 (1993); J.K. Burdett and S.L. Price, *Phys. Rev. B*, **25**, 5778 (1982).
4. P.P. Edwards and M.J. Sienko, *Acc. Chem. Res.*, **15**, 87 (1982).
5. J.J. Gilman, *Czech, J. Phys.*, **45**, 913 (1995).
6. J.J. Gilman, *Phil. Mag. B*, **67**, 207 (1993).
7. W.A. Harrison, *Electronic Structure and the Properties of Solids*, W.H. Freeman, San Francisco, 1980, p. 193ff.
8. J.J. Gilman, *Phi. Mag. B*, **71**, 1057 (1995).
9. J.J. Gilman, in *Metal–Insulator Transitions Revisited*, P.P. Edwards and C.N.R. Rao, Eds., Taylor and Francis, London, 1995, p. 269ff.
10. W.L. Faust, *Science*, **245**, 37 (1989).
11. A.V. Belik, V.A. Potemkin and N.S. Zefirov, *Dokl. Akad. Nauk. SSSR*, **308**, 882 (1989).
12. For example, see D.A. McQuarrie, *Statisical Mechanics*, Harper and Row, New York, 1976, p. 67.
13. W. Yang, R.G. Parr and T. Uytterhoeven, *Phys. Chem. Mineral*, **15**, 191 (1987).
14. R.Y. Goble and S.D. Scott, *Can. Mineral.*, **23**, 273 (1985).
15. J.N. Plendl, S.S. Mitra and P.J. Gielisse, *Phys. Stat. Sol.*, **12**, 367 (1965).
16. R.G. Pearson, *J. Mol. Struct. (Theochem.)*, **300**, 519 (1993).
17. L. Pauling and J. Waser, *J. Chem. Phys.*, **18**, 747 (1950).
18. R.G. Pearson, *J. Phys. Chem.*, **98**, 1989 (1994).
19. N. Dass and M. Kumari, *Phys. Stat. Sol.*, **127**, 103 (1985).
20. In the notation of Pauling and Waser, $f = k$ and $-6g = k'$.
21. G. Simmons and H. Wang, *Single Crystal Elastic Constants*, MIT Press, Cambridge, MA, 1971.
22. C.F. Kittel, *Introduction to Solid State Physics*, 3rd Edn., John Wiley and Sons, New York, 1967, p. 183ff.
23. R.G. Pearson, *Int. J. Quantum Chem.*, **56**, 211 (1995).
24. I.C. Sanchez, J. Chu and W.J. Chen, *Macromolecules*, **26**, 4234 (1993).
25. I.C. Sanchez, *J. Phys. Chem.*, **97**, 6120 (1993).
26. M.H. Cohen, M.V. Ganduglia-Pirovano and J. Kudrnovsky, *J. Chem. Phys.*, **101**, 8988 (1994); idem, ibid., **103**, 3543 (1995).
27. B.G. Baekelandt, G.O.A. Janssens, W.J. Mortier, H. Toufar and R.A. Schoonheydt, *J. Phys. Chem.*, **99**, 9784 (1995).
28. A. Cedilli, B.G. Baekelandt and R.G. Parr, *J. Chem. Phys.*, **103**, 8548 (1995).
29. J.L. Dunham, *Phys. Rev.*, **41**, 713, 721 (1932).
30. For a good discussion, see I.N. Levine, *Molecular Spectroscopy*, Wiley–Interscience, New York, 1975, Chapter 4.
31. D. Steele, E.R. Lippincott and J.T. Vanderslice, *Rev. Mod. Phys.*, **34**, 239 (1962).
32. K.P. Huber and G. Herzberg, *Constants of Diatomic Molecules*, Van Nostrand Reinhold, New York, 1979.
33. J.H. Rose, J.R. Smith and J. Ferrante, *Phys. Rev. B*, **28**, 1835 (1983); A. Banerjea and J.R. Smith, ibid., **37**, 6632 (1988).
34. J.L. Graves and R.G. Parr, *Phys. Rev. A*, **31**, 1 (1985).

Index

Acids, Table, 3
Activation hardness, 82,
Adiabatic I and A, 34
Adsorption, 167
Alloys, 161
Ambident bases, 7, 21, 85
Anions, Table, 8
Anomeric effect, 15
Anti-symbiosis, 14
Aromatic compounds, Table, 80, 166
Atoms, Table, 51

Bases, Table, 4
Bloch functions, 126, 141
Born charging energies, 12, 153
Born Mayer equation, 129, 182
Brillouin zone, 139, 141

Catalysis, 23, 167, 169
Cations, Table, 7, 35
Charge capacity, 170
Cohesive energies, Table, 1, 133, 149, 158
Conductivity, 143, 147
Contact potential, 146
Correlation function, 107
Crystal orbitals, 141, 157

Density of states, 167
Dielectric constant, 150, 153, 176
Diels–Alder reaction, 72
Dislocations, 176
Divide and conquer method, 125
Dunham formulation, 192

Elastic moduli, 175, 179
Electrochemistry, 90, 126, 145
Electron-hole pair, 152
Electronic spectra, 39, 44, Table, 41
Electron transmission spectroscopy, 34
Energy gaps, Table, 149
Exchange-correlation energy, 30
Exciton, 152, 155
Explosives, 179
Extended Huckel theory, 167

Fermi distribution, 144, 147
Fermi energy, 138, 143, 147, 159
Fluctuations, 106, 184
Fluctuation–dissipation theorem, 105, 107
Free electron model, 137
Free radicals, Table, 74
Frontier orbitals, 38, 47, 68
Fukui function, 42, 46, 81, 84
Fukui function, nuclear, 190
Fuzzy logic, 25

Glide activation energies, 17
Grand potential, 48, 106, 113, 120
Grain boundaries, 170
Grand canonical ensemble, 105, 115
Gruneisen constant, 188

Harmmond principle, 29
Hardness functional, 113, 122
Hardness of molecules, Table 36, 193
HSAB Principle, 3, 48, 169
Huckel theory, 82, 104, 115, 138, 160, 178

Information entropy, 119
Intrinsic strength, 4
Ionic radii, 128, 132, 135

Jahn–Teller effect, 112

Kink motion, 176
Kohn–Sham orbitals, 30, 56
Koopmans theorem, 38

Linear free energy relation, 18
Local hardness, 43
Local softness, 43, 48, 167, 169

Magic numbers, 163
Marcus equation, 21
MEP, 87
Metal hydrides, 9, 23
Metallization, 176, 180
Methyl cation affinity, 12
Moh scale, 175, 181
Morse equation, 131
Mulliken EN, 33, 161
Mulliken population analysis, 87
Murnaghan equation, 186

Nuclear softness, 190
Nucleofugality, 18
Nucleophilic reactivity, 16

Octet rule, 104
Olefins, reactions, 64, 66, 72, 75

Particle in box, 89,119, 137
Pauling EN, 10, 24, 49, 54, 78
Penetration error, 131
Perturbation theory, 44, 46, 150
Photoelectric threshold, 144
Physical hardness, Table, 182
Plastic deformation, 175, 181
Polarizability, 3, 39, 47, 88, 92, 176
Polymerization, 73
Promotion energy, 10, 67
Pseudopotential, 89

Radius-ratio rule, 128, 132, 135
Relaxation time, 107
REPE, 45, 166, 180
Rotational isomerization 113

Shear stress, 175, 179
Shell structure, 103, 163
Solubility, 22, 93
Solvation energies, 10, 153
STM, 168
Symbiosis, 14
Symmetry coordinates, 110

Tight-bonding model, 127, 138, 160
Thermal expansion, 185, 187
Thomas–Fermi atom, 29
Transition state method, 55
Transition states, 69, 70, 76, 112, 177

UBER, 194
Uncertainty Principle, 118

Variance, 109, 118
Vertical I and A, 34
Vickens test, 175, 177

Wade–Mingos rule, 104
Wave vector, 137
Work function, 165, 151, 160, 163

Xα method, 29, 125